PLANT
PROPAGATION
IN PICTURES

Evergreen azaleas can be propagated easily from cuttings taken from the plant in early summer.

PLANT PROPAGATION IN PICTURES

Montague Free

*How to increase the number of
plants in your home and garden by
division, grafting, layering, cuttings,
bulbs and tubers, sowing seeds and spores.*

ILLUSTRATED STEP BY STEP WITH MORE THAN
THREE HUNDRED AND FIFTY PHOTOGRAPHS

Revised and Edited by
Marjorie J. Dietz

DOUBLEDAY & COMPANY, INC.,
GARDEN CITY, NEW YORK

For J. H. BEALE and H. E. DOWNER
my classmates at the Royal Botanic Gardens, Kew, England.

Books by Montague Free

All About House Plants
All About African Violets
Plant Propagation in Pictures

Library of Congress Cataloging in Publication Data

Free, Montague, 1885–1965.
 Plant propagation in pictures.

 1. Plant propagation. I. Dietz, Marjorie J.
II. Title.
SB119.F8 1979 635.9′1′53
ISBN 0-385-12986-6
Library of Congress Catalog Card Number 76–56290

PREFACE TO THE REVISED EDITION

In his preface for the first edition of *Plant Propagation in Pictures*, Montague Free wrote:

> *This book is for amateur gardeners. It is designed to aid those who have no special facilities for plant propagation but who are interested in the processes involved and who may want to increase their stocks of certain favorite plants. Although it is primarily a "picture book," it is desirable to read the Foreword and also the introduction to each chapter.*

This revised edition is still such a book—and what a book it is! What a pleasure to see such carefully detailed and planned photographs, many of which Montague Free took himself or set up for the photographer (often with the aid of his wife Maude, who contributed much of her own knowledge and experience). And what a relief it is to see gardening procedures and techniques done right, and explained and demonstrated by one who believed he should practice first what he was to write about later.

Two techniques demonstrated with superb style (see pp. 35 and 36) show the correct way to make a shallow drill for fairly small seeds and a deeper one for larger seeds. These are simple procedures—I have made such rows hundreds of times in my own vegetable garden, yet no one ever showed me that keeping one foot on the line would prevent the row from ending in a bow shape rather than being straight as an arrow. Styles in gardening shoes (!) may change, but these and other basic gardening techniques, as presented by a master gardener such as Montague Free, are for all seasons and all generations.

Marjorie J. Dietz

CONTENTS

FOREWORD TO REVISED EDITION

The propagation of plants. The most intimate friend of my boy-hood happened to be the son of the head gardener of a large private estate, and through him I became interested in the mysteries of plant propagation at a very early age. He showed me that it was possible to root the side shoots removed from tomatoes when pruning them to a single stem; and that young apple trees could be produced by planting the "pips" (seeds) from the apples we shared. In one re-spect both of these bits of information were misleading. Tomatoes are not usually propagated by cuttings, because it is more advanta-geous to raise them from seeds; and apples are not raised from seeds, excepting understocks for grafting and when the production of new varieties is desired, because the characteristics of the parent may not be reproduced in the seedlings. On the other hand, the information so casually acquired led to the knowledge that there are at least two methods whereby plants can be propagated—sexually, by seeds and spores; and asexually by vegetative parts of the plant.

For over fifty years I was concerned, more or less, with various aspects of plant propagation and at divers times held the position of propagator in three extensive establishments. But, even with long experience, the miracle of the germinating seed, the rooting of cut-tings, and the successful union by grafting of two diverse plants, never lost its power to provide the thrill of accomplishment. This little bit of autobiography is introduced to indicate the enduring in-terest conferred by a knowledge of plant propagation.

Everyone who maintains a garden practices plant propagation to a certain extent, even though it may be no more elaborate than scattering seeds of annuals in the places where they are expected to bloom. But, sometimes, raising plants from seeds is not so simple as this, for certain seeds require special handling to ensure satisfactory germination. This may involve: the removal of the pulpy covering, cracking, filing, abrading, or otherwise treating the seed coat to allow the moisture to penetrate to the embryo; stratification in a moist medium to enable the seeds to retain their viability; storing in low temperatures to bring about certain changes in the seed essential to germination; the provision of high or low temperatures according to the subject; the right light conditions and, in some cases, excision of the embryo. A knowledge of the various factors influencing germination is of immense value to the gardener whether in the backyard or professional category.

There are some plants which do not come true from seeds. A knowledge of the various ways of vegetative propagation enables us to increase our stock of these plants by division, cuttings, grafting, or budding, methods which as a rule ensure the perpetuation of the desirable characteristics (and also, unfortunately, the undesirable ones) of the parent plant.

It may be maintained that these specialized methods belong in the province of the professional—the nurseryman and the commercial grower. This is true, in part, but, as will be seen later, there are many occasions when it is desirable for the home gardener to do his own propagating, and knowing how is essential.

In nature plants propagate themselves sexually by seeds and spores; asexually (vegetatively) by bulbils, bulblets, corms, cormels, tubers, rhizomes, runners, offsets, suckers, stolons, layers, divisions, and cuttings.

It may be asked how on earth plants can make cuttings of themselves, and it must be admitted that the term is not literally accurate, but in effect it is true enough. The twigs of some willows (which form roots with greater ease than those of any other tree) are so brittle that they become detached as a result of windstorms or by birds lighting on them. Furthermore some species of willow prune themselves annually by developing what is known as an abscission layer at the base of some of the twigs. (An abscission layer is a zone of corky cells similar to that which enables the leaves of deciduous trees to fall off in the autumn.) As almost everyone knows, the natural habitat of many willows is along riverbanks and some of these broken-off twigs fall into the water, are carried downstream, and some, becoming lodged on a mudbank, take root, and thus a new tree is started.   M.F.

1 SEEDS AND SPORES

Seeds are infinite in variety, in size, and in shape. Some are infinitesimally small: such as begonia, one ounce containing about a million seeds; in contrast the double coconut, the largest known seed, may weigh upward of fifteen pounds. In shape they vary from the spherical, shotlike seeds of canna to the curious angular seeds of the brazil nut, and the winged seeds characteristic of the catalpa and the empress tree (*Paulownia*).

In texture they may be soft, as in the cases of ismene and crinum, whose seeds may be easily picked to pieces by a fingernail, or extremely hard as in the ivory-nut palm, which has been used as a substitute for ivory.

The leaves of seedlings of some plants are very different from those of mature plants. In some cases, if these seedlings are propagated by cuttings before the adult foliage is formed, the juvenile habit is retained. An example of this is to be found in juvenile forms of false-cypress (*Chamaecyparis*) and arbor-vitae (*Thuja*).

The primary purpose of seeds is to ensure the perpetuation of the species and in many cases Mother Nature has devised means whereby they are widely distributed. There are some which have *parachutes*, such as the dandelion, milkweed, and salsify; there are *gliders*, such as maple, pine, birch, and empress tree; *floaters in water*, which include sea coconut or double coconut and water-lily; there are *aerial floaters*, which have cottony appendages such as do the seeds of willow and poplar; other air-borne seeds include many of the bromeliads and orchids; then there are the *hitchhikers*, such as burdock, beggars tick, and the interesting unicorn-plant. The fruit of the last-named has a long curved horn, which, when mature, splits into two curved grappling hooks. Several of these are produced in a ring surrounding the fruiting stems, which lie upon the ground, so that when a horse or similar animal steps in the middle of a clump, the hooks become engaged by the hair of the fetlocks and thus may be distributed over a large territory. Squirrels and birds (bluejays in particular), in their endeavors to provide food for themselves during the long winter, also are responsible for planting seeds of oak, walnut, and hickory when they fail, as they often do, to find the seeds which they have so carefully hidden. Sunflowers, also, may be inadvertently distributed by birds. At my home, "Buttonwoods," every year sunflowers came up in odd places, doubtless from seeds dropped by chickadees or nuthatches after they had visited the bird feeder.

You, too, probably help to disseminate seeds when you come in contact with such weeds as burdock, beggars tick, or galium, whose hooked fruits or seeds cling to clothing as well as to the hair of animals. There are some whose fruits are *explosive*, such as squirting cucumber, witch hazel, and legumes such as Scotch broom. Then there are the *sprinklers*, such as those of the poppy and the East Indian lotus. There are many fruits which are attractive to birds, such as those of the honeysuckle, holly, and turquoise berry. These have pulpy coverings which are disgested by the birds and the hard seeds are evacuated, unharmed and with a modicum of fertilizer.

Seeds

More flowering and vegetable plants including culinary and ornamental herbs are produced from seeds than from any other means. Usually seed germination presents no problem to the gardener, provided that the seeds are given their minimum requirements of air, moisture, and sufficiently high temperature.

These requirements are met in outdoor planting by covering the seeds with soil deep enough to keep them moist, but not so deep that they do not get sufficient air, or so deep that the food stored in the seed is exhausted before the growing shoot reaches the surface.

Temperature requirements are met either by starting the seeds in a greenhouse or in the home (if you have a well-lighted window or fluorescent light setup in a cool room—60° to 70° F.), in a hotbed or a cold frame; or by deferring planting until the ground is warmed up sufficiently in the spring.

The tomato, eggplant, and pepper are vegetables whose seeds require warmth for germination and subsequent growth as well as a long growing period before their fruit can be harvested. Both of these requirements are met in the North by starting the seeds indoors in early spring—about 6–8 weeks before killing frosts end outdoors. A popular flowering annual whose seeds require similar treatment is the petunia.

Many seeds succeed best in cool temperatures and some can advantageously be planted outdoors in late fall. These will germinate in the early spring when the temperature is right for them. Some of the *hardy annual* seeds, such as love-in-a-mist, annual larkspur, California-poppy, and the Shirley poppy, may be planted in late summer or early in the fall so that they germinate and form young plants which are capable of surviving the winter. These plants usually bloom early and last longer than those from spring-sown seeds.

The sweet pea is still another well-known flowering annual whose seeds must be sown in late fall or very early spring. In the vegetable garden, the sweet pea's close relative, the garden pea, also requires early sowing. St. Patrick's Day is the traditional date for sowing garden peas over much of the North. Among other vegetables that benefit from an early start are the radish, the broad or fava bean, curly and upland cress, onion, and the herb parsley.

Biennials such as hollyhock, English daisy, Canterbury bells and cup-and-saucer bells (*Campanula medium* and *C. medium calycanthema*), foxglove, forget-me-not (*Myosotis*), honesty (*Lunaria annua*), Greek mullein (*Verbascum olympicum*), sweet William, pansy, and wallflower (*Cheiranthus cheiri*) are sown from midspring to midsummer.

Midspring to late spring (late April or early May in the North) is the best time for sowing most of the *hardy perennials*, although the early bloomers which ripen their seeds before August can, to advantage, be sown as soon as they are ripe. The seeds may be sown in a prepared seed bed in the open, although it is preferable to have some kind of transparent or translucent cover to avoid the danger of heavy rain.

Except for those perennials with a mixed ancestry, as a result of efforts of the plant breeders, which cannot be relied on to come true, raising from seeds affords the gardener a ready means of quickly increasing his stock. The named varieties of some genera such as phlox, peony, chrysanthemum, when grown from seeds, usually produce progeny mixed in character, with the chance that most of them will be inferior to the parent. (On the other hand, there is a bare possibility that some of the offspring may be better.) Among the plants which can be satisfactorily raised by seeds are the following: species of columbine such as *Aquilegia caerulea* and *A. chrysantha*, assuming that there are no other columbines nearby, which could be mixed with them as a result of the ministrations of bees or hummingbirds; plume-poppy (*Macleaya cordata*), coreopsis,

Cattleya trianaei—capsule and ripe seeds, probably exceeding a million in number.

bleeding-heart, gas-plant (*Dictamnus*), Christmas-rose (*Helle-borus*), gay feather (*Liatris*), perennial flax (*Linum*), Lobelia, bal-loon-flower (*Platycodon*), and such perennial herbs as sage, chives, lavender, the mints, and thyme.

Here is a member of the pineap-ple family whose capsules have just opened to display the seeds waiting for a puff of wind to blow them away. This is an epi-phyte, or "tree percher." Those seeds that lodge on a moss-covered trunk or branch are likely to grow.

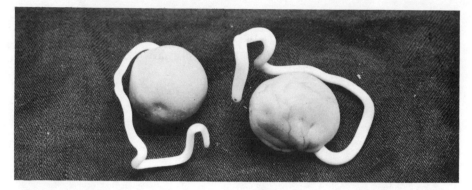

The soft fleshy seeds of ismene germinate when they are ready, re-
gardless of moisture. The seeds pictured here started to grow in September while in an ash tray on the author's desk.

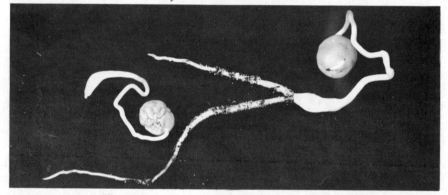

April 3, following year. On the right is a seed that germinated and
made roots when placed on moist soil in a terrarium. The one to the
left was kept in the ash tray and is living on its own "fat." Notice that
the seed is wrinkled owing to the withdrawl of nutrients.

May 6. The seed has disappeared. The plant has made a small bulb and
the beginnings of a leaf. It should be planted outdoors in fairly rich
soil as soon as danger of frost is past.

Unicorn-plant (*Proboscidea*), showing the vicious hooks which may become tangled in the hair above the hoof of a horse who happens to step on it. In this way the seeds may be distributed over an area covering several miles. Incidentally, I read somewhere that the owners of horses in "Proboscidea countries" detest this plant because of the injury to horses which may be brought about by the constant irritation of these grappling hooks.

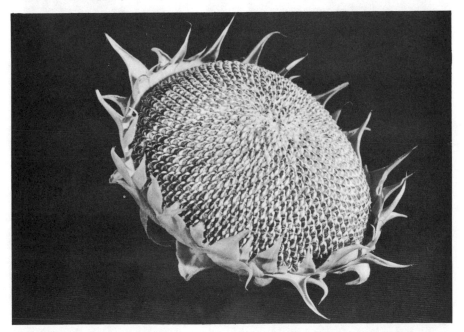

Sunflower seeds are favorites of many seed-eating birds. In gardens everywhere sunflowers germinate annually as a result of seeds which nuthatches, bluejays, and chickadees inadvertently drop.

Fruit of magnolia in various stages. Notice the seeds, which have a pulpy covering. These are eaten by birds, the pulpy covering is digested, and the seeds are evacuated unharmed.

Salt-shaker-like fruits of the Oriental poppy. Seeds are spilled out of the holes beneath the cap by action of wind or by an animal brushing against the stems when they are dry and brittle.

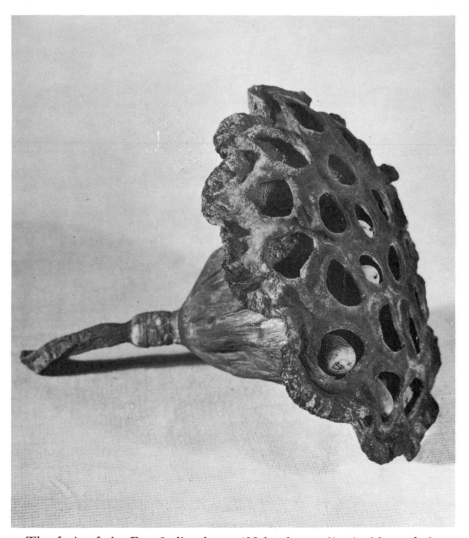

The fruit of the East Indian lotus (*Nelumbo nucifera*). Most of the seeds have already fallen out from this specimen. These drop into the water and may float for a considerable distance before they sink into mud at the bottom of the pond.

The seeds of avocado germinate no matter whether the seeds are planted with the pointed end up (the right way), on their sides, or upside down.

Sowing seeds indoors

Seeds require a porous and well-drained soil for best results, especially when they are sown indoors. A standard soil mixture for seeds in general consists of equal parts of loam (garden soil), leafmold and/or peat moss, and sand. Here you see a flower pot with drainage in the bottom and equal parts of sand, rotted compost, and peat moss suitable for *Saintpaulia* (African-violet), begonia, gloxinia, etc., which thrive on it. In the spoon there is a small quantity of pulverized limestone to neutralize the acidity of the peat moss. The strainer with ⅛-inch mesh is used to provide an extra-fine layer on the top of the seed pot. It can also be used as an aid in covering the seeds when this is necessary. For more information on soil mixtures as well as soilless mixes and the use of sphagnum moss, see Chapter 10.

Preparing the soil. A good way to ensure that the soil is not too loose or packed too tightly is to fill the pot with moist soil, not wet, not dry, tap it on a bench or table, strike off surplus soil so that it is level with the pot rim, and then press it down, or firm it, about a half inch below the rim. Firming can be done by means of a tamper like the one pictured (left), which was made from a section of a tree branch about 2 inches in diameter, or you can use the bottom of a drinking glass.

Under no circumstances should you "puddle" the soil. The pot on the right demonstrates the bad effect of "puddling." Corn was seeded in both of these pots and received exactly the same treatment with measured amounts of water. The only difference was that the soil in the pot which shows no growth whatever was puddled by stirring it when it was wet, thus driving out air. This explains the often-reiterated advice: "Do not attempt to work on the soil when it is wet."

Sowing the seeds. No attempt should be made to cover small seeds such as those of begonia and African-violet (*Saintpaulia*); merely gently press them into the surface with a tamper.

The pot can be covered with glass or plastic film or slipped inside a kitchen plastic bag, which should be removed as soon as germination is evident.

A white-seed variety of lettuce was chosen to illustrate the ideal spacing for seeds of this size.

Because lettuce seeds are supposed, under some conditions, to require light for successful germination, these were covered with a very thin layer of sand. Ordinarily the rule to follow when planting seeds indoors is to cover them with soil equal in depth to twice the diameter of the seeds.

There are two methods of moistening the soil. One is to stand the pot in an inch or two of water and leave it there until moisture shows on the surface.

Another way is to use a fine spray.

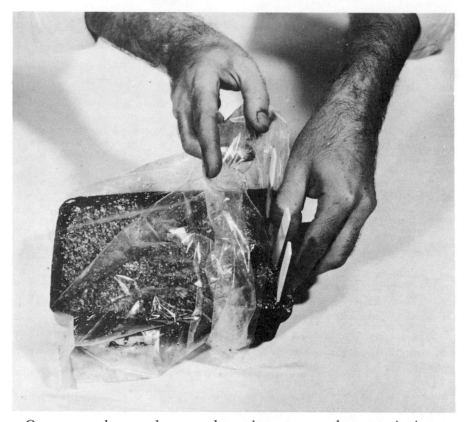

Once watered, a good way to keep the pot or seed tray moist is to slip it into a plastic bag. Keep in a warm place (but out of direct sun). As soon as the seeds germinate, open the bag or remove entirely. Place the tray or pot in a sunny window or under fluorescent tubes. If the seedlings seem to be "stretching," move the pot or tray closer to the tubes.

One means of germinating small seeds, such as begonia, gloxinia, and African-violet, is in square Mason jars. Enough finely screened soil is put in a jar to bring it to the top of the shoulder and is pressed down with a knife blade. Or use a plastic food or small storage box—make a few holes for ventilation in the lid with a "hot" ice pick or nail.

There are two ways of introducing the seeds into the jar. One is by means of a knife blade on which the seeds are placed.

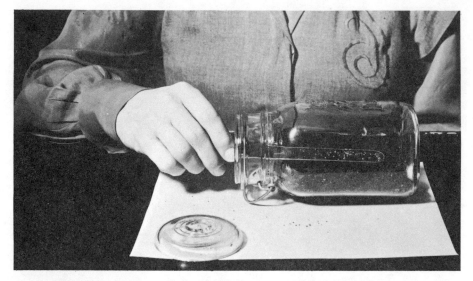

The blade with the seeds is carefully inserted into the Mason jar and by tapping it gently against the side of the jar the seeds are evenly distributed.

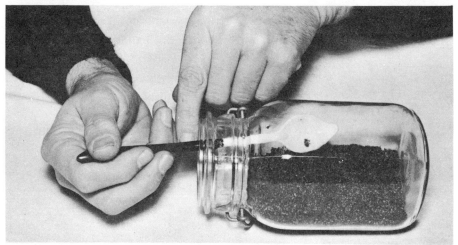

Another and perhaps simpler way is to put the seeds in an iced-tea spoon, holding the handle in one hand and lightly tapping the spoon with the finger of the other hand. After the seeds are sown, the top is then put on the jar without any rubber ring. This will largely prevent loss of water by evaporation and at the same time permit interchange of air. The only advantage these methods have over the way described on page 23 is that, if the soil is sufficiently moist when it is put in the jar, it will remain that way until seeds have germinated and are large enough to be transplanted.

When the seeds have germinated and are large enough to be pricked out (transplanted), they can be removed with the aid of a teaspoon.

Handling small seedlings such as these is facilitated by the use of a "transplanting fork," which can be made from a 6-inch wooden pot label by cutting the pointed end to a sliver and then making a notch in it.

Although the orthodox soil mixture in which to start seeds is equal parts of sand, garden soil, leafmold and/or peat moss (right), there are some gardeners who prefer a sterile medium such as vermiculite (left).

It is necessary, however, when a sterile medium is used, to provide nutrients for the seedlings. This is the same flat of seedlings after two applications of a soluble commercial fertilizer to the vermiculite section.

Rebutia minuscula (Echinocactus minusculus). Flat of seedlings in orderly array.

Seedlings of *Trichocereus spachianus*. Before the young plants become too crowded in the seed pot or flat, they should be transplanted.

Here are seedlings of a culinary herb sweet marjoram, a tender perennial in most northern regions that is usually treated as an annual. The seeds have been sown in "pinches" and are now ready to be transplanted about 3 inches apart in flats.

Marigold seedlings growing in peat pots. Notice how the roots grow through the walls of the pot. Both plant and pot are set out in the ground. Not shown are the handy peat pellet pots in which one or two seeds are sown, eliminating further transplanting.

Orchids

The seed-sowing methods shown in the preceding pages can be used for most house or greenhouse plants—with the exception of one large group, orchids.

Raising orchids from seeds used to be a gamble (some will say that it still is!), up to 1903, when Noel Bernard demonstrated that mycorhiza fungi were necessary to ensure the development of orchid seeds. This was a great improvement over the haphazard method that prevailed previously. About twenty years later Dr. Lewis Knudson of Cornell University showed that it was possible to germinate sterilized orchid seeds in flasks, on a sterilized nutrient solution of agar; thus proving that fungi were *not* essential to germination. Essential are: (1) use of a nutrient solution containing cane sugar plus mineral salts; (2) great care to do the work under aseptic conditions.

If you are interested in raising orchids from seeds, information on this complicated process can be found in *Home Orchid Growing* by Rebecca T. Northen (3rd revised edition, 1970, Van Nostrand-Reinhold Co., New York) and *How to Grow Orchids* by the Sunset Editorial Staff (Lane Books, Menlo Park, California).

From seed pod to flowering plant.

Sowing seeds outdoors

One of the first requisites for sowing seeds outdoors is the preparation of a seed bed. This is accomplished by first digging up the soil, by hand if the area is small or by a mechanical tiller for larger areas. Soil lumps should be broken up, then the area should be raked to remove stones and trash. When very small seeds are to be sown, such as those of poppy or portulaca, they are sown directly on the raked surface, which is then patted down with the end of the rake. See page 36. Seeds may be broadcast; or sown in rows in drills or furrows first outlined with a line, as shown below.

Most vegetables (peas, beans, beets, carrots, corn, squash, cucumber) are sown in long rows since they are likely to remain in place without transplanting. Short rows are best for most flowering plants such as marigolds, pansies, and zinnias which may be destined for transplanting to other sections of the garden. Many gardeners also like to sow short rows of lettuce, broccoli, cabbage, cauliflower, and such herbs as sweet basil, borage, chervil, coriander, dill, fennel, and sweet marjoram. Or the seedlings can be thinned and left to mature in the short rows, since they are usually moved to other parts of the garden.

If the soil contains so much clay that it is heavy and sticky, avoid planting seeds when it is wet, as this may result in puddling the soil to such an extent that air is driven out and the germination of the seeds is prevented. An exceptionally sandy soil may require the addition of organic matter to increase its moisture-holding capacity. Peat moss, rotted manures, compost, and leafmold are useful for this purpose.

The soil in a small area, such as that enclosed by a cold frame or in a section outlined as a temporary site for seed-sowing, can be easily modified. If the soil is a heavy clay, add a 3-inch layer of sand and an equal amount of peat moss or sifted leafmold. If the soil is sandy, add a double amount of peat moss or leafmold. These materials should be dug into and thoroughly mixed with the upper 3–5 inches of the existing soil. This should be done several weeks ahead of planting in order to give the soil a chance to settle. Or, provided it is not too wet, it can be consolidated by tramping.

In order to make a shallow drill a rake is used, keeping one foot on the line to ensure that the drill is straight.

Seeds may be sown directly from the packet. It is essential to sow them evenly and thinly. The seeds of sweet-alyssum should be spaced about 1 inch apart, best accomplished by "thinning" the seedlings.

The seeds are covered by raking across the drill, and then the soil is firmed down by patting it with the head of the rake.

The best tool for making a furrow for reception of larger seeds is a draw hoe. Notice that the right foot of the operator is kept on the line to prevent it from being pushed out of position, which would result in a curved row.

Seeds are sown thinly or thickly according to the subject and the objectives in view. In the case of vegetables, such as peas, that are destined to mature without any transplanting, the aim is to have an ultimate space of about 2 inches between the plants. Therefore seeds are sown about 1 inch apart to take care of loss of seeds by failure to germinate. On the other hand, if the seedlings are to be transplanted, spacing may be a little closer.

Covering the large seeds may be done as shown on page 36, but some prefer to do the covering with their feet by shuffling along the row in such a way that the seeds are covered and pressed down at the same time.

Trees and shrubs from seeds

Those trees and shrubs which have comparatively large seeds, such as ash, oak, linden, or euonymus, can be planted in the open ground just as soon are they are available. Give them a covering of soil equal in depth to two or three times the diameter of the seed.

Those species of maple which ripen their seeds early in the year, also willow and poplar, are short-lived and must be sown as soon as they are ripe. Chestnut, oak, and walnut lose their vitality if they are allowed to become thoroughly air-dry. Therefore, these also must be sown or stratified as soon as they are ripe. Stratification is the name applied to the practice of putting seeds in layers (strata) in a moist medium such as peat moss, sand, sawdust, or a mixture of these. It is usually considered more desirable to mix the seeds throughout the medium, but the name still stands.

Seeds which have fleshy coverings, such as those of apple, cotoneaster, barberry, dogwood, holly, and rose, should have the pulp removed before they are planted.

There is a group known as "two-year seeds," which ordinarily do not germinate for two years or more after planting. They include some of the roses, hawthorns, and cotoneasters. A convenient way of handling these, together with those mentioned above, is to put them, mixed with peat moss, in small jars which are covered either with glass or with plastic film such as is used in the freezer. The seeds should be placed in such a way that some of them are visible from the outside. The jars should be put in the refrigerator and looked at from time to time. As soon as any signs of germination are apparent, the seeds should be removed and planted in a regular soil mixture of equal parts loam, sand, and peat moss or sifted leafmold.

Small seeds, such as those of hemlock, spruce, fir, larch, and other conifers together with mock-orange (*Philadelphus*) and spirea, may be kept dry, preferably under cool conditions (50° to 60° F.), during the winter and sown in the spring. These usually can be handled most effectively by sowing them in either flats (shallow boxes about 3 inches deep) of any convenient size, usually about 12 by 18 inches, or small pots.

Once the seeds have been sown, it is necessary to keep the medium moist. This can be done by means of a fine spray or by standing the pots or flats, temporarily, in water about 1 inch deep, which will be absorbed by capillarity. Just as soon as moisture appears on the surface, they should be removed from the water.

Except for shade-loving plants, when the seedlings are up they should be given all the sun available.

Rhododendrons, including azaleas and other acid-soil plants, such as the red-veined enkianthus, leucothoe, heath (*Erica*) and heather (*Calluna*), also have small seeds that germinate best when sown fresh—from late fall through early spring—indoors. Sphagnum moss (see page 59) alone is the ideal growing medium, but good results have been obtained with mixtures of sphagnum moss and perlite, peat moss and sand, and leafmold and sand. If you live near bogs, you can collect your own sphagnum moss; otherwise it is available at garden centers or from mail-order seed companies. The moss can be used live or dried, but should be shredded into small pieces and rubbed through a sieve. Soak it thoroughly, then squeeze out excess water. Fill small seed pans (or any containers that can be covered with panes of glass or plastic) with the moss, pressing it down to obtain a final layer about 2–4 inches thick. Useful containers for this purpose can be found in hardware or variety stores: look for bread or shoe boxes with clear plastic lids or plastic food containers of varying sizes. The seeds are sparingly scattered over the moss surface with no attempt being made to cover or water them in. Then the container is covered with its lid or glass or plastic to maintain the necessary moisture. Under average room temperatures, the seedlings will begin to emerge within a few weeks, at which time the container can be placed under fluorescent lights. When the first true leaves have formed, the little seedlings can be transplanted into a humusy mixture. Wean them gradually from their enclosed environment and continue to grow under artificial light until warm weather arrives and they can be moved outdoors.

Berried trees and shrubs

Usually it is desirable to remove the pulpy covering of certain seeds such as holly, viburnum, and cotoneaster before stratifying or sowing them. This can be done by putting the berries in water and leaving them for a few days until covering is soft and easy to remove.

The strawberry-like fruits of Kousa dogwood are being soaked in water to soften the pulpy covering.

Here the seeds are in process of being removed from the fruits with the aid of a household strainer.

The seeds were put in moist peat moss on November 2 and placed in the refrigerator, which was kept at approximately 40° F. The white arrow is pointing toward two of the seeds.

On January 31 they were removed from the refrigerator and placed on the shelf in my study. Signs of germination were evident on February 11. They were photographed on February 13, after which they were transplanted.

Holly berries at left; seeds with pulp removed at right. There are usually about four seeds in each berry.

These can be sown in wooden flats, first treated with a preservative such as copper naphthenate (Cuprinol), either broadcast (left) or in rows (right).

The seeds are then covered with about 1 inch of soil, sand, or, as here, with a layer of vermiculite.

Spores

Although as far as the gardener is concerned there is little difference between seeds and spores, botanically they are vastly different. A *seed* is the result of the fertilization of an egg cell in the ovule by its union with the sperm cell from a pollen grain. When mature it contains a plant embryo surrounded by seed coats. A *spore* in many cases, as in ferns, is a specialized cell which under suitable conditions germinates to form a prothallus, which on its underside produces male elements (antheridia) and female elements (archegonia). Spermatozoids produced in the antheridia swim by means of vibrating cilia in the film of moisture usually present on the underside of the prothallus, and unite with the egg cells produced in the archegonia. The result is the fertilized egg cell (oöspore), which is perhaps more nearly comparable to a seed. The fertilized egg cell goes through a process of segmentation which finally leads to the development of a plant similar in appearance to the fern which produced the spore.

Asplenium ebenoides (Scott's spleenwort), a hybrid between the walking fern (*Camptosorus rhizophyllus*) and the ebony spleenwort (*Asplenium platyneuron*), possesses some of the characteristics of both parents—its fronds tend to root at the tip, as do those of the walking fern; and they are partly pinnate, showing the influence of the spleenwort. The parentage of the hybrid was suspected for a long time and finally was demonstrated by Margaret Slosson—presumably by sowing spores of both species in the same germinating medium. In Alabama it is more abundant than elsewhere, and is believed to be self-perpetuating there.

We don't hear a great deal about fern hybrids, even though they are not uncommon. They may occur when two species are growing in close enough proximity to enable their spores to be distributed and mixed in the same area. If the sperm of one species happens to wander into the archegonium of another species, a hybrid may result.

It is easy to raise ferns from spores. All that is necessary is to "first catch your hare" or, in other words, first get your spores. These usually are produced on the undersides of the fronds. They require close watching, using a hand lens, if necessary, to see when the spores are ripe, which occurs when the spore cases begin to split open. When this is seen, the frond should be removed and placed on a piece of white paper. By the following morning, in all probability, the paper will be discolored by millions of spores.

A good way to plant these is to secure first a common brick, on the upper surface of which should be spread a layer of sand and sifted leafmold or peat moss. This can be put in the oven for about one half hour at 250° to sterilize it. Carefully remove the brick and put it in a vessel containing about 2 inches of water. A plastic bread box serves admirably for this purpose. As soon as the soil is obviously moist, spores can be scattered on the surface. This method can also be adapted to sowing very small seeds, such as those of begonia, gloxinia, and saintpaulia.

Though happy in the limestones of Bartholomew's Cobble (Sheffield, Mass.) and similar locations, the charming little walking fern (*Camptosorus rhizophyllus*), so named because of its lengthened frond tips, which root to form new plants, is difficult to establish, often simply walking into oblivion.

Ebony spleenwort (*Asplenium platyneuron*) is a delightful little fern, easy to transplant into the shaded rock garden. It and the walking fern are the parents of *Asplenium ebenoides*, a true bigeneric hybrid, which occurs in nature only infrequently.

Scott's spleenwort (*Asplenium ebenoides*) is intermediate between the two parents, with fronds that have a tendency to form plantlets at the tips; notice one at top left-hand corner that has started "walking."

Fern prothallia and seedlings of African-violet (*Saintpaulia*) on a brick in a plastic box that was kept enclosed to retain a moist atmosphere. If you are wondering why the saintpaulia were mixed in with ferns, it is because, as my wife always insisted, I was a pessimist. You see, the fern spores were planted first and nothing happened for a long time, so I decided that they were not viable and sowed saintpaulia seeds, which promptly germinated along with the spores.

2 PROPAGATION BY DIVISION

Vegetative or asexual propagation is a term applied to the propagation of plants from parts other than seeds or spores. They may consist of stems, leaves, roots, or, rarely, of fruits (prickly-pear opuntia). Propagation may be effected by division of the rootstock, by cuttings, or by graftage. In general, plants raised by asexual propagation reproduce the parent plant exactly, but there are a few exceptions to the rule. If leaf cuttings are made of the snake-plant—*Sansevieria trifasciata laurentii*, the variety whose leaves are marked with yellow longitudinal stripes—the progeny lacks the yellow stripes, possibly because they revert to the original species. Some varieties of *Pelargonium* (house geranium), notably an old variety known as 'Mrs. Gordon', when propagated from stem cuttings, have white flowers tinted with pink, but, when raised from root cuttings, they have deep red flowers. A similar phenomenon is seen in the case of a variety of *Bouvardia* known as 'Bridesmaid', which has pink flowers when propagation is by stem cuttings, but, when raised from root cuttings, the flowers are deep red. One cannot always rely on *Saintpaulia* (African-violet) varieties to come true from leaf cuttings. Sometimes three distinct color forms, "blue," white, and pink may originate from a single leaf.

The understocks used in budding and grafting also may exert an influence on the size of the varieties which are grafted on them. Malling 9, for example, is well known as a dwarfing understock for apple trees.

During recent years important discoveries pertaining to propagation as well as changes in techniques have taken place. Perhaps the most astounding discovery is meristem propagation or clonal multiplication, which was developed by Georges Morel of France. This is a complicated laboratory technique in which many new plants are developed from cell tissues taken from the growing tips of the plant's branches—the new plants all being identical to the parent, perhaps not an appropriate term under the circumstances. Several practical benefits have resulted from this procedure, an important one being the production of virus-free plants, including chrysanthemum, carnation, lily, potato, and strawberry.

Meristem propagation has revolutionized the orchid world by increasing the number of named plants (it used to take years to increase the stock of one named orchid by division) and having them available much sooner and at reasonable prices. These mericlones, as such orchids are called, as well as easier potting methods (see page 57), are important factors in the increasing popularity of orchids as house plants.

Polyethylene, a plastic film permeable to air but retentive of moisture, has been put to good use in air-layering and grafting and also as a covering for cuttings and seeds. Plastic food bags come in handy for temporary storage of cuttings; or the cuttings can be rooted in the bags if a mixture of moist peat moss and sand or some other rooting medium is added.

As for cuttings themselves, there is the matter of wounding the stems by cutting a slit through the bark.

Constant mist, a technique by which wilting is prevented by keeping the leaves almost constantly moist under a mistlike spray, is a valuable means of propagating certain plants.

Division

Division is among the surest and easiest methods of plant propagation. It is applied almost exclusively to herbaceous plants—especially the hardy kinds. The most important thing to remember is to do the job at the right time. This occurs when the plants are dormant, in spring or fall. A good general rule to follow is to divide early-blooming plants in the fall and late-blooming ones, such as hardy asters and chrysanthemums in the spring. There are a few exceptions: primroses can be divided immediately after the flowers fade, in the spring before growth starts, or early in the fall; the bearded irises summer to early fall; Oriental poppies and madonna lilies should be separated when the foliage becomes yellow in late summer.

Greenhouse and house plants are best propagated by division, when possible, in the spring, when their new growth is about to begin.

When only a few plants are needed, divisions can be set in the garden where they are to remain, with care taken, however, to ensure that the soil does not become dry. If a large number of plants is desired, as, for example, when Japanese irises are separated into single shoots (which involves considerable root injury), the divisons should be set in a special plant bed in which the soil has been made porous by the addition of sand and peat moss. It may be necessary to shade them from bright sun until they begin to grow and the soil should be kept moist but not soggy.

House plants

In all cases involving severe root disturbance it is desirable that the pot should be well drained. The usual way of doing this is to put a piece of broken pot over the hole in the bottom and then about ½ inch of smaller pieces.

A plant of African-violet (*Saintpaulia*) in need of division.

Five rooted divisions were obtained from this specimen.

This shows a three-crown African-violet plant in process of being divided.

The divisions are potted in 3-inch pots.

Two pots of mother-of-thousands or strawberry-geranium (*Saxifraga stolonifera*). The one at the left is a plant with several crowns, one at the right is a "single-crown" plant. (See also Chapter 5.)

This plant has been turned out of its pot and is in process of being divided.

In this picture you see the results. It won't be long before they look like the single-crown plant in the rear.

Ferns such as this (*Polystichum tsus-simense*) often become rootbound, so division is necessary. Start the process by cutting into the root ball with a knife.

Then separate the crowns, as is shown here. When the division is made for propagation purposes, they can be divided into many more by cutting or pulling them into pieces, each having a single crown.

In this case only two plants were desired.

Here is a snake-plant (see also page 126) capable of being divided to make several plants.

A good way to do this is to stick two hand forks back to back into the root ball to pry it apart.

In this case four for one were obtained.

Aspidistra, sometimes known as parlor-palm or cast-iron-plant, can be handled similarly to the snake-plant. The one you see here could be divided into several plants, each one with a leaf and a growth shoot.

Most orchids can be propagated by division by utilizing the "back bulbs." Often it is desirable to separate them a year ahead of the time when the actual division is done, by cutting through the rhizome. This is a much slower process than meristem propagation (see page 48) but is a practical method for home gardeners.

Here you see a new shoot originating from the back bulbs of *Oncidium tigrinum*. This orchid is potted in osmunda fiber, the chopped-up roots of a fern (*Osmunda* spp.), a medium that is being replaced by redwood and fir barks, alone or mixed with other materials. The barks are easier to use and the ferns are facing extinction in many regions.

The time to do the dividing is soon after the flowers fade and when the new growth is about to start.

The "makings" needed for potting orchid divisions are shown here. A recommended mixture for cattleyas is 3 parts medium-size fir bark (top), 1 part shredded sphagnum moss (left), and a handful of perlite (right).

A *Cattleya* hybrid in need of repotting has been knocked out of its pot. The arrow points to approximate area for cutting the plant into two divisions with a sharp, sterile knife. The divisions are then gently pulled about and any dead roots are carefully cut off.

Old potting material and dead roots are put in newspaper and destroyed.

This is the right size pot for this division and should allow about two years of growth.

The plastic pot is prepared for planting by arranging old broken pieces of clay pots (crock) and a layer of coarse, washed gravel so they aid rather than hinder drainage.

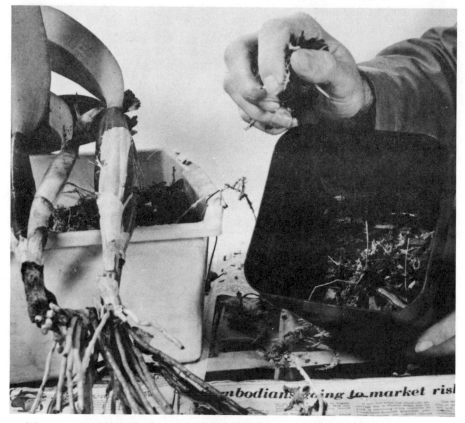

Next a 1-inch layer of the potting medium is placed over the crock.

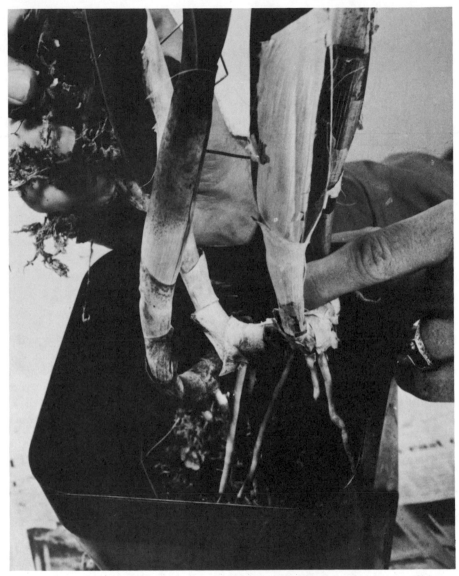

Place the division with the cut end toward the "back" of the pot. This, in most cases, enables the growing point to be centered. Holding the plant firmly, pack the growing medium between the roots and sides of the pot.

Cattleyas should be potted firmly; use an aluminum potting stick and strong thumbs to build up the medium so the division finally rests on the surface just below the rim of the pot.

The pseudobulbs can be tied to a stake to steady them in their new surroundings. Newly divided orchids should be watered sparingly until new growth appears. (The stake has a rubber cap for eye safety.)

Hardy herbaceous perennials

Coreopsis verticillata (golden shower) being dug up preparatory to division.

A good way to start is by prying the clump apart with two spading forks placed back to back.

The hole to receive the division is made with a spade and should be made large enough to contain the white shoots and the roots without crowding.

The fibrous root system of day-lily (*Hemerocallis*) requires a similar technique but more muscle! A clump is dug up preparatory to dividing.

Pushing through the clump two spading forks, back to back and close together to pry apart the rootstock, is preferable to chopping the clump with a spade.

Further division into pieces of planting size is done with a hand fork. In some cases they can be pulled apart with the hands. The divisions should be reset at the same depth as they were before digging them up.

Ordinarily this plant of hosta, or plantain-lily, would be considered a good size for planting to give the best garden display.

When you are thinking in terms of propagation, the plant can be divided as seen here.

Here is a clump of lily-of-the-valley dug up in the fall and divided as shown under *Coreopsis verticillata.*

Pieces of suitable size ready for planting. If many plants are needed, they could be separated into individual shoots with roots attached.

The first step in propagating peonies by division is to dig up the entire plant and shake the soil away from the thick, fleshy roots. If the soil happens to be sticky and clayey, it may be necessary to wash it off with a hose.

Some three- to five-eyed divisions can be pulled off with your hands. Others must be cut apart.

Here the clump has been divided into three- to five-eyed pieces. One of these divisions has only one eye. Small pieces such as this are desirable only when division is done primarily to increase the number of plants.

When planting the divisions it is essential that the soil be firmly tamped to prevent future subsidence. None of the eyes is more than 2 inches below the surface.

When the peony division is set at the right depth, fine topsoil is put in between and over the roots so as to be in full contact with them. Tramping will then pack the soil firmly to eliminate all the subsurface air pockets. Be sure to cut off the stubs of the stalks to lessen the danger of attack by botrytis disease.

A division dug up about six or eight weeks after planting shows the new roots, the white fibrous ones, which have already started to grow.

Early fall is the best time to divide peonies because the soil is warm enough to encourage formation of new roots, which anchors the plants and lessens the danger of "heaving" as a result of alternate freezing and thawing during the months ahead.

When dividing phlox, dig up the entire plant and pry strong pieces from the outer edges of the clump with a hand fork. The worn-out center (note the thin spindling stems) should not be used for propagation purposes unless absolutely necessary.

The divisions should be planted in holes large enough to contain the roots without crowding. Cover them with soil and pack it firmly around them, either with the hands, as shown here, or with the feet, as shown under peony on page 71.

Primroses can be divided as soon as the flowers have faded, provided that the divisions can be planted in a partly shaded position and can be watered during dry spells. If these conditions cannot be met, it is better to defer the dividing until the fall or early the following spring. In this picture you see a plant being dug up.

The clump is easily divided by pulling apart the crowns.

The divisions are planted by making holes with a trowel.

Three plants obtained by separating have been planted and are being watered. Notice that the flower stalks have been removed to divert the energy of the plant from seed production.

3 PROPAGATION BY CUTTINGS

Stem cuttings of house plants

Although these can be rooted indoors at almost any time, the summer months are preferred for most of them because young plants of good size are developed before it is necessary to bring them indoors before frost. Another advantage is that the propagation can be done outdoors.

The simplest way to raise house plants from cuttings is merely to insert them in soil outdoors in a shady spot. Usually early August is the best time to do this. Here you see iresine (also known as bloodleaf), patience-plant, and begonia.

Campanula isophylla, the Ligurian harebell (sometimes called star-of-Bethlehem), is often grown as a house plant, although when suitably located outdoors it can endure temperatures down to zero. It can be propagated by cuttings and, when grown as a house plant, needs a cool temperature. Add about a trowelful of limestone to each 6-inch pot of soil.

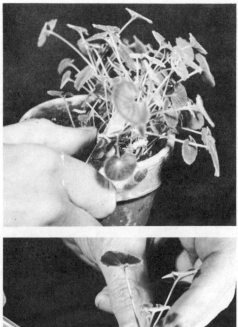

Cuttings of the Ligurian hare-bell should be made in the spring, when it is possible to obtain short, stocky growth from the base of the plant.

The cutting is inserted in the standard sand, soil, and peat moss mixture (1 part sand, 1 part peat moss, 1 part garden soil), surfaced in this case with vermiculite. If more convenient, sand can be used instead.

Cutting is placed in a plastic bag. Ordinarily it is not necessary to close the top of the bag, but under very dry air conditions it is better to do so.

Eight weeks later the cutting is a well-rooted plant.

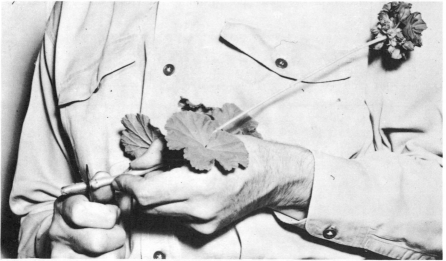

This shows the method of making the basal cut of the house geranium. Notice the position of the thumb of the right hand, which supports the cutting without the danger of slicing into it. Be careful, as accidents can happen, as my wife and daughter would readily attest to!

A 3-inch pot is filled with standard seedling and cutting mixture, a hole is made with the blunt end of a pencil, and a little sand is poured into the bottom of the hole. This provides the free drainage which is desirable.

The cutting is then inserted and the soil is pressed down firmly around about it.

Cuttings of (left to right, top to bottom) fuchsia, abutilon, patience-plant, wandering Jew, and coleus.

These can be inserted in a bulb pan filled with standard soil mixture or a mixture of peat moss and sand or any of the packaged soilless mixes available at garden centers. In this picture the method of making the soil firm about the base of the cutting is shown.

A shoot of lantana prepared for insertion. The undesirable portions, leaves, immature seed heads, and flowers encircle the cutting.

Lantana—the result obtained from the preceding.

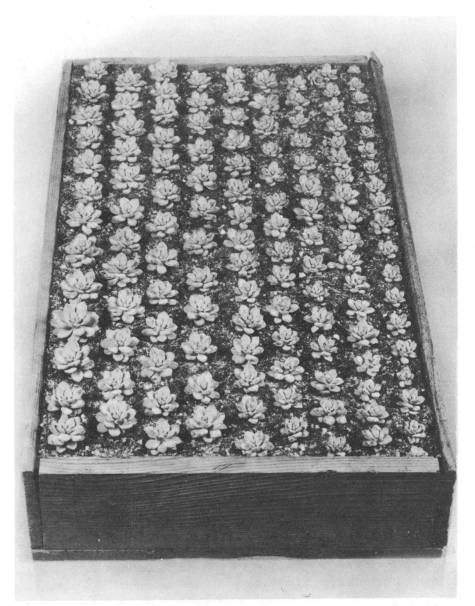

Stem cuttings of *Echeveria microcalyx*. These can be inserted directly in a flat containing the standard mixture. They do not need any covering.

Poinsettia. Commercial growers propagate poinsettias for Christmas sales by cuttings. The cuttings are often rooted in Jiffy 7 peat-pellet pots, in which case both cutting and peat pot are simply planted into a permanent clay or plastic pot where flowering will occur. Those who receive these poinsettias during the holidays can rebloom the plants by cutting back the stalks in spring after the colored bracts and flower clusters have faded or fallen. Keep the plants in a sunny location all summer and feed a few times with a fertilizer recommended for house plants. A convenient way to handle the plants is to sink the pot outdoors in a sunny spot, remembering to water if the soil becomes dry. Prune the new growth once or twice to keep the plants compact. If you wish to propagate your poinsettia, use these early summer prunings as tip cuttings. Do not prune the plants after early August.

The cuttings of poinsettia, about 6 inches long, may be inserted in 4- or 5-inch pots filled with the standard 1-1-1 mixture or one of the soilless mixes. The top of the plastic bag is rolled down to facilitate the insertion of the pot. Then it is rolled up and closed at the top with a tie-on. Or each cutting can be rooted in a Jiffy 7 peat-pellet and then transferred into a permanent pot.

You can expect something like this five or six months later.

Three scrawny house plants, left to right, impatiens, coleus, and vinca. They have no claim to beauty in their present condition.

Here they are again, with the material suitable for cutting removed. Beauty is still conspicuous by its absence and it would be better to discard them. In the foreground there are cuttings at the left of each plant; leaves, debris, and flowers at right.

This is a "Forsyth" cutting pot. It consists of a 6-inch bulb pan in which a small porous clay pot is centered. The drainage hole of this pot is plugged either with cork or chewing gum. This pot is kept filled with water. After the initial watering the seepage from the central pot usually is sufficient to keep the rooting medium moist.

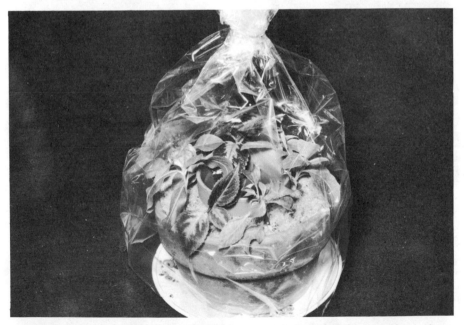

In the dry air of the average living room it is helpful to enclose the potted plant within a plastic "tent." In fact the plastic bag is so efficient at maintaining moisture that the "Forsyth" method is hardly necessary!

Multiple-crown plant of *Saintpaulia* (African-violet). Plants with single crowns usually are preferable.

The surplus crowns can easily be removed by cutting them with a sharp knife—a kitchen paring knife will serve.

Four crowns have been taken from the original plant. The wounds are dusted with a fungicide. The "duster" is a small piece of lamb's wool attached to a straightened-out paper clip. The severed crowns can be planted in the standard cutting mixture. They should receive extra shade for a 'week or two, by which time they should be able to withstand the normal light for *Saintpaulia*.

Often diseases can be by-passed by rooting cuttings of non-affected parts. One such is the crown rot which sometimes affects the African-violet.

The affected part has been cut away.

A good way to root crowns such as this is to put them in a saucer of water. After a few weeks enough roots are formed so that transplanting is necessary.

One of the difficulties encountered when transferring water-rooted cuttings to soil is that the roots cling together when they are removed from water.

This can be overcome by putting the new container in a pan of water, which enables the roots to spread out naturally.

Pour in the soil while the roots are still submerged. The soil should be filled in level with the pot rim and the plant taken from the pan of water. The soil can be made sufficiently firm merely by tapping the pot bottom on a bench or table.

Screw-pine (*Pandanus*) is an excellent house plant when it is young and has an undivided crown, but the time usually arrives when it becomes too large for its location and is cluttered with surplus shoots. These can be taken off, the cuts made as near as possible to the parent stem.

The so-called "bedding plants," some of which have already been discussed (house geranium and lantana), can be included here. Alternanthera (which can also be divided), echeveria, and lobelia are examples.

About one third of the top growth should be removed by shortening the leaves.

The trimmed shoot is inserted in sand in a 3-inch pot with the rooting medium packed firmly around the base of the shoot.

This is then covered with a small glass jar or tumbler; or the pot can be inserted in a plastic food bag.

Cuttings of hardy shrubs and trees
GENERAL CARE

A humid atmosphere must be maintained about the cuttings, which means that the cold frame or propagating case should be kept closed most of the time. It is a good plan to ventilate the frame by tipping the sash for a half hour or so every morning, and on hot, humid days a crack of ventilation may be left all day. For a propagating box that can virtually take care of itself, see the polyethylene-enclosed flat—the "covered wagon"—on page 138. The rooting medium (see pages 101 and 226 for specific recommendations) must be kept constantly moist. Bottom heat is desirable but not necessary and can be easily supplied by simple heating cables with built-in thermostats offered by mail-order seed and nursery companies. Or it is possible to buy or build more elaborate propagating boxes equipped with heat and fluorescent tubes. See page 237 for such a case.

Shade from bright sunshine is necessary, but ample light is essential. The frame may be placed in the shade of a building or a large tree. In the open I have had good success by whitewashing the underside of the glass and providing an additional covering, for use when the sun is bright, of a double thickness of cheesecloth attached to light wooden strips cut to fit the frame; or lath shades could be used. Indoors, enclosed propagating boxes must be kept out of direct sun but can be kept under the "cool" light of fluorescent tubes.

Yellowing and fallen leaves should be removed daily. When the cuttings have rooted (in three to nine weeks or longer), they should be dug up and potted or planted out in another frame, where they must be shaded for ten days or until they become established in their new quarters. They can be transplanted to the open the following spring.

Timing

A start can be made with softwood cuttings of *Neillia sinensis* in mid-May, continuing with lilac at the end of the month. However, June and July are the big months for this kind of propagation. Lilac cuttings, for example, should be put in soon after the flowers have faded in late May or early June; some, as you will see if you consult the Propagation Notebook in the Appendix, root very well from mid-June insertions; but you can take a chance with most shrubs in July, including evergreen azalea, barberry, boxwood, forsythia, daphne, heaths and heathers, hydrangea, tamarisk, spirea, viburnum. Some, such as evergreen barberries, false-cypress (*Chamaecyparis*), juniper, mahonia, pieris, rose, thuja, and viburnum, are likely to give a better percentage of rooting if they are inserted toward the end of July or early in August. Don't be deterred from trying cuttings of any kind if you feel like it, but cryptomeria, most kinds of magnolia, and large-leaf evergreen rhododendrons can be disappointing, although polyethylene-enclosed flats and rooting hormones have improved results with rhododendrons.

The differences between softwood, half-ripened wood, and mature wood cannot be defined with any accuracy. Softwood cuttings are those taken early in the growing year before the growth is finished for the season. Mature wood is that at the end of the season of growth, usually in September and October in the North. The half-ripened wood is that which comes in between, say in July and August.

Softwood cuttings of shrubs and trees

When making softwood cuttings it is desirable to prevent them from wilting. If you have a variety of cuttings which have to be collected from a considerable distance, wilting can be prevented by carrying them in a pail lined with wet newspaper. Or use plastic bags from the kitchen.

A variety of cuttings ready for insertion. Left to right, top row: wisteria, rose, lilac, boxwood; bottom row: bush honeysuckle, pieris, garland butterfly-bush (*Buddleia alternifolia*), and winter jasmine (*Jasminum nudiflorum*).

Two sections of cold frame are being prepared for reception of the cuttings. Underdrainage is important. If the soil does not naturally drain, it is necessary to excavate to a depth of 6 inches or so and fill it with coarse sand or gravel. The dark-colored medium is a mixture of sand and peat moss. On the right is coarse sand.

The sand and, to a lesser extent, the peat moss and sand should be packed down firmly. Here a half brick is being used.

With a strip of builder's lath to serve as a guide the cuttings are inserted and made firm, with particular attention paid to the base of the cutting by pushing the dibble (planting stick) in from the side.

When two or three rows have been put in, they are thoroughly watered, which helps still further to consolidate the sand.

The cold frame sash is then put in place. Shade is necessary to prevent the temperature from rising too high. This can be done by painting the underside of the glass with whitewash or by putting on a covering of cheesecloth, double or triple thickness, or both. It may be necessary to ventilate by raising the sash on hot, humid days.

This is a homemade propagating frame with a window sash cover.

When only a few cuttings are to be inserted of each kind, usually it is better to put them in pots.

It is desirable to put extra drainage material in the bottom of the pot.

The big advantage of inserting the cuttings in pots for hardening off is the easy removal of those which root ahead of the others.

Summer cuttings of wintercreeper (*Euonymus fortunei vegetus*). These easy-to-root plants can be started by inserting them in a mixture of sand and peat moss. They should be kept in a shaded cold frame with the sash on until they are rooted. Roots are produced more freely in some media than in others. These two groups of cuttings were inserted at the same time, the batch on the left in a 50-50 mixture of peat moss and sand, while those on the right were rooted in a mixture of equal parts of sand, good garden soil, and leafmold.

Cuttings with strong roots should be potted up immediately. Here a cutting is being potted individually in a 3-inch pot in a mixture composed of equal parts of soil, sand, and leafmold. Be sure to provide proper drainage at the bottom and to water the cutting thoroughly after potting it up.

Those to the left (*Euonymus fortunei vegetus*) were potted singly late in summer. Those to the right, including buddleia, beauty-bush (*Kolkwitzia*), *Symphoricarpos*, and *E. kiautschovicus*, were left in the original pots. All were carried over winter in the cold frame and plunged to the rim in sand. After a few hard frosts they were covered with a deep blanket of hardwood leaves and the sash was put in the frame.

The following spring the pots were taken from the frame to plant in the open ground. First, a shallow trench was made against a line to mark the row, then a trowel was used to plant the potted-up cuttings.

Here cuttings are being set out directly from the cutting pot into a trench made with a spade against the line. The roots are slipped into the trench and the soil is pushed firmly against them with the fingers. Be sure to water well. Leave ample space between cuttings so both root and top growth can continue without crowding.

Three years later *E. kiautschovicus* has attained a stature of 3 feet.

Similarly, garland butterfly-bush (*Buddleia alternifolia*), beauty-bush (*Kolkwitzia amabilis*), and *Symphoricarpos* x *chenaultii* (above) range from 3 to 4 feet.

This specimen, now thoroughly established, will maintain its annual rate of quick growth until it matures, when the rate will drop.

Some of the best-loved native plants are reputedly difficult to propagate. Among these is the trailing-arbutus (*Epigaea repens*), which, however, gave almost 100 per cent rooting from cuttings inserted during July in two successive years. Equally good results were obtained from shoots of the current season and those which were cut to include an inch or two of wood of the preceding season.

A flat is filled with a 50-50 moist mixture of sand and peat moss. A slit trench is made by pressing in a ruler. The cutting in the foreground is of the current season's growth, the one to the rear has about 1 inch of the wood of the preceding year.

The cuttings are inserted more or less horizontally with the objective of having the underside of the leaves as close as possible to the rooting medium.

When all cuttings were inserted, the soil surface was covered with a layer of unmilled sphagnum moss, except for one row which was left uncovered as an experiment. The roots were not quite so heavily produced as in the moss-covered section. Another method that uses sphagnum moss is to wrap a few strands of the moist moss around the stems of the cuttings before inserting in the rooting medium.

The flat was then covered with a pane of glass to increase the humidity in the vicinity of the cuttings and to reduce the need for watering. The glass covering was tilted about 1 inch with the lower edge ideally falling ½ inch short of the side wall of the flat; thus rain falling on the glass ran down into the rooting medium, through which it was dispersed by capillarity. This was done to avoid as far as possible the use of our well water, which was alkaline—pH 7.5.

The following spring the cuttings were well rooted and the most difficult part of the operation to be done was that of transplanting them without too much loss into an acid, humusy soil mixture.

One fairly certain method to avoid any root disturbance whatever is by planting the cuttings in a container, buried so that the rim is a little bit below the surrounding soil. If this is not convenient or if you want plants to give away to friends, they can be potted in 3½-inch pots, which preferably should be kept in a shaded and partly closed cold frame for a few weeks.

Wintertime cuttings

Many plants, especially the narrow-leaved evergreens, can be rooted from cuttings which are made during October, November, and December.

There are two methods of making cuttings of arborvitae. One is to cut off a small branch with a sharp knife.

Another is to pull off a small branch—this is known as taking a cutting with a heel.

Here are samples of various cone-bearing evergreens to indicate the size of cuttings preferred—the labels are 6 inches long. There are two types of cuttings, the left one of each pair is made with a heel.

Often the use of a root-inducing growth regulator is helpful. The method of application is to take a half-dozen or so prepared cuttings, dip the butts in a saucer of water, shake off the surplus, and then dip them into a container of the root-inducing powder.

A flat filled with sand, the cuttings closely inserted in rows, and made firm by tamping the sand.

Cuttings are given a thorough soaking to help settle further the sand and the flat is then put in a cold frame and left there all winter. In order to protect them from severe freezing they can be given a light covering of excelsior or salt-marsh hay. The sash is put on and during severe weather it is covered with a layer of salt-marsh hay. If it is not convenient to put them in a cold frame, they could be put, as these were, on a glass-enclosed, unheated porch until late in November, when there was danger of their freezing. They were then brought into my study and placed on the floor and kept there until April, when they were taken back to the porch.

A homemade propagating case. This may be a box, 12 by 9 by 5 inches, with a transparent superstructure which consists of two pieces of glass 12 inches wide by 10 inches high for the sides and two pieces 8¾ inches wide by 10 inches high to form the ends. These are put in place before the rooting medium, which helps to hold them upright. The corners at the top can be fastened together with strips of Scotch tape. The roof is a piece of glass 13 by 9½ inches which rests on the side walls. This propagating case was kept all winter in the bay window of my study. This was a little too warm for best results, even though the temperature dropped at times to 45° F. Ideally they should be placed in a cold greenhouse where the temperature is maintained between 45° and 50° F.

Here are samples taken on May 26 showing the roots produced in vermiculite. These, except for the juniper at the left, are ready either to be potted or to be "lined out" in the open ground.

A similar lot taken from the flat of sand. Although roots of these are not so far advanced as those above, they ultimately came along and about 80 per cent rooted.

Hardwood cuttings of deciduous shrubs

Many deciduous shrubs can be propagated by means of hardwood cuttings which are taken any time during the winter, but preferably as early as possible after the leaves are fallen.

Cuttings are made of shoots which grew during the preceding growing season. Some of them are satisfactory only when the cutting contains a terminal bud, including rose-of-Sharon (*Hibiscus syriacus*) and weigela. With the majority of the shrubs suitable for propagation in this way the entire length of the shoot can be used, cutting it into pieces 6 to 10 inches long, preferably with at least three nodes, or "joints." When cuttings are made in this way, it is important that the proximal ends should point in the same direction. They then are tied into bundles and are either buried in a sandy knoll where the drainage is perfect or are put into layers in boxes and covered with moist sand, peat moss, or sawdust. Ideally they should be exposed to a temperature of 65° or 70° F. for four or five weeks to initiate the formation of callus, after which the temperature should be lowered to 35° to 40° F. Your choice of either of these two methods you use depends on whether you have suitable storage space, such as a garage or unheated cellar where the low temperatures prevail.

In the spring the cuttings should be planted outdoors, with only the terminal bud showing above the surface, set about 6 inches apart in rows 2 feet apart.

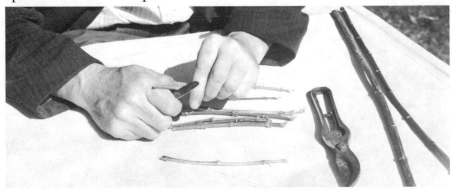

Cuttings are made in lengths of from 6 to 10 inches, preferably with at least three nodes. They can be cut with either a knife or pruning shears, making the cut a quarter to a half inch below the joint.

Those species which root easily may be cut with sharp shears. With more difficult ones it is better to use a sharp knife.

When the cuttings are made, they are tied into bundles of twenty-five or less, each one labeled with the name of the plant. With the butts of the cuttings all pointing in the same direction the cuttings are then covered with moist sand or peat moss or sawdust and kept in a cold cellar during the winter. The temperature should be from 65° to 70° F. for four or five weeks, then it should be reduced to 40° F.

A bundle of weigela cuttings removed from sand storage. Note the callus especially on the center cutting. There is too much top growth at this stage for best results.

Cuttings in water

This is a favorite method of beginners but it has a disadvantage in that the roots tend to clump together when they are removed from the water and it is difficult to spread them naturally when potting them. It can be done, however, by following the method described on page 91.

Boxwood (*Buxus sempervirens* 'Rotundifolia'). At the left are new shoots put in during July; only one rooted. At the right are cuttings of older wood made in April of the same year, three of nine well rooted, four starting, two no signs of roots. They were kept on the shelf where they stood still until the summer of the following year, when this photograph was taken. Obviously, this is not a good method in this particular case.

Cuttings of the golden weeping willow. These are nearly ready for planting outdoors; larger cuttings could have been used.

Growth made by dwarf arctic willow (*Salix purpurea* 'Gracilis'), from cuttings started in water in the spring. Photograph was made in the early fall of the same year.

Notice the excellent root system.

Cutting of Michaelmas daisy or New York aster (*Aster novibelgii*), which took to the water. The propagation of hardy asters in this way is a possible means of by-passing a wilt disease that affects them.

Cutting of oleander: A few pieces of charcoal to absorb obnoxious gasses are recommended by garden writers, but it is doubtful that it has much effect on rooting one way or the other.

Cuttings of *Saintpaulia* (African-violet) rooted in water. (See also page 91.)

Leaf cuttings

Leaf cuttings offer the gardener a ready means whereby some plants can be propagated. Ordinarily the plants originated this way are replicas of the parents, but, as noted below, there are some exceptions.

There are numerous plants which are viviparous (producing buds which are capable of regeneration), such as pickaback-plant (*Tolmiea menziesii*) and some varieties of water-lily which produce plantlets on the leaves. Another example is the plant variously known as sprouting-leaf, miracle-leaf, air-plant, life-plant, good-luck-leaf, and floppers (*Kalanchoe pinnata*).

Pickaback-plant—when growing in the wild, the outer leaves of the pickaback-plant ultimately topple over to the ground, where the plantlets on the leaves take root and grow. In propagating them under artificial conditions they can be cut off with 1½-inch leafstalks and inserted into individual flower pots in the standard 1-1-1 mixture. Keep them shaded for a few weeks.

Many tropical water-lilies have viviparous leaves. The one shown here is Dauben water-lily (*Nymphaea* x *daubeniana*). Notice the plantlets that surround the main plant. Before cold weather comes in the fall, these leaves may be cut off and each one put in a water-holding container at least 4 inches deep and 1 foot or more across. This should be filled with soil to within 1 inch of the top, in which the plantlet is placed. Then the container should be filled to the brim with water. Keep it in a sunny window or in a greenhouse with a minimum temperature of about 60° F. until it has warmed up enough outdoors to warrant placing it out, usually about the first week in June in the vicinity of New York City. If more than one or two plants are needed, a tank of some kind is necessary. In this case the plantlets should be potted in 4-inch pots which are submerged about 3 inches below the surface of the water.

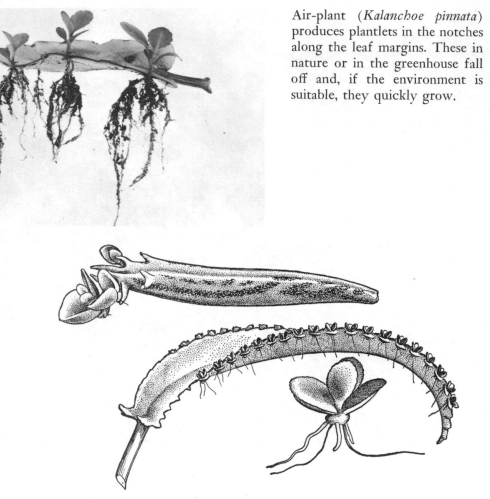

Air-plant (*Kalanchoe pinnata*) produces plantlets in the notches along the leaf margins. These in nature or in the greenhouse fall off and, if the environment is suitable, they quickly grow.

K. daigremontiana and *K. verticillata* behave similarly, except that the plantlets develop only at the leaf tips on the latter.

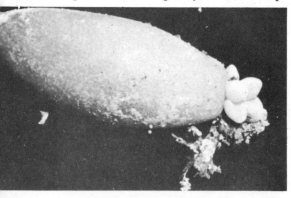

Sedum morganianum (donkey's tail) and others also propagate themselves in the same way.

Adult plant of donkey's tail hangs in a sunny window with geraniums.

Almost any of the subtropical species of *Sedum, Echeveria,* etc., can be propagated by pulling off leaves and sticking their bases in a flat of the 1-1-1 mixture or of plain vermiculite. Leaf cuttings in the foreground are of mother-of-pearl or ghost plant (*Graptopetalum* [syn. *Sedum*] *paraguayense*).

Sedum species with leaf cuttings at right.

Peperamia obtusifolia, with leaf cuttings in various stages.

Pot containing leaf cuttings of *Saintpaulia* (African-violet). These were inserted on May 18, photographed October 25 the same year. The precocious one in bloom is 'Sir Launcelot'.

Rooted plants, more than one per leaf. The original leaf can be replanted if you want more plants. These usually look like the parent, but occasionally a single leaf may go haywire and produce plants that bear either pink, white, or purple flowers, or produce different kinds of leaves. These are known as mutations.

A leaf cutting of *Saintpaulia* rooted in water. This method is preferable to the one frequently recommended of covering a jar with waxed paper in which holes are punched, through which the leaf stalks are pushed into the water. The leaf also benefits from the extra humidity conferred by the partial enclosure.

A leaf of rex begonia is cut with 1-inch leafstalk (petiole) and the main veins severed. The stalk is inserted in the sand so that the entire leaf lies flat on the rooting medium. It may be necessary to put two or three stones on the leaf to hold it down, or pin it down with bent wire sections or hairpins.

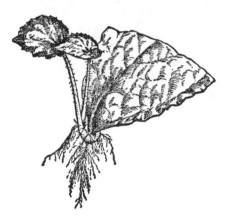

If preferred, the leaf can be cut to wedge-shaped pieces, each containing a main vein.

Leaf-bud cuttings

Toward the close of the 1930s, Dr. Henry T. Skinner worked on leaf-bud cuttings of rhododendrons and obtained excellent results. Such cuttings, which are also successful with camellias and a few other plants, are usually made in July and after being treated with a growth regulator (root-inducing chemical) are inserted in a mixture of sand and peat moss in a propagating box or enclosed flat.

Snake-plant (*Sansevieria trifasciata laurentii*)—leaves cut into pieces 3 to 6 inches long root readily, but the characteristic yellow stripes of the variety *laurentii* are lacking in the progeny.

The varietal response to this method varies considerably and many kinds of rhododendrons prove to be difficult and cantankerous. Propagating rhododendrons by stem cuttings is much preferred by both home gardeners and nurserymen today, especially since the use of polyethylene (see page 135) has created such a favorable environment for this propagation method. However, leaf-bud cuttings have their place, especially for the hybridizer or home gardener who wishes to increase the numbers of a particular plant. It is also possible to save leaf-bud cuttings from the trimmings of stem cuttings—a thrifty practice not without appeal.

Leaf-bud cutting of *Rhododendron decorum*.

Left, root development of *R. decorum* a few weeks later. Right, root ball and shoot about five months after cutting was made.

Branch of rhododendron showing kind of shoots from which buds may be cut.

Leaf-bud cuttings of *R. maximum*.

Cuttings made and inserted in flat with constant mist (see page 142).

Root cuttings

One advantage to be gained from increasing plants by root cuttings is that little is needed in the way of skill or equipment to be successful. Grafted varieties, of course, should not be propagated by root cuttings because the understock is different from the top and, for garden purposes, usually inferior. But for others that are amenable (see lists below) it is an effective way of propagation.

In the fall dig up the roots of the variety selected for propagation. If it is herbaceous and a number of new plants are desired, the entire plant may be dug up to supply a sufficient number of roots. If, however, only a few new plants are needed, enough roots can be obtained by pushing in a spade alongside the clump and prying outward. In the case of trees and shrubs the best method is usually to dig into the soil about midway between the outermost spread of the branches and the center of the bush or tree to get the fairly young, vigorous roots. Be very sure that the roots you dig up belong to the plant you wish to propagate!

The thickness of the desirable roots varies according to the species. With phlox, for example, the stringlike roots—about $\frac{1}{10}$ inch in diameter—not the whiskery ones—are suitable; those of Japanese anemone are $\frac{1}{8}$ inch or a little larger; while in the case of woody plants they may be up to $\frac{1}{2}$ inch across.

Before insertion the roots are cut into lengths of about 2 to 6 inches and are equally spaced in rows about 2 inches apart on the surface of the rooting medium made up of a mixture of sand, soil, and peat moss in equal parts, placed in a flat, and pressed down fairly firmly to 1 inch below the rim. They are then covered with $\frac{1}{2}$ to 1 inch of sand and watered thoroughly. When only a few plants are required, several kinds can be put in labeled rows in each flat. During the winter the flats are kept in a cold frame and after a few frosts are mulched with light, littery material such as oak leaves or excelsior and left until spring.

In the spring shoots will appear above ground and, when they have attained a height of about 3 inches, the plantlets may be transplanted to a nursery row or other convenient place outdoors and shaded for a few weeks. Some of the late-flowering herbaceous plants may give a few blooms the first season if growing conditions are good, but don't expect them to be at their best until the following year.

Among woody plants these are listed as capable of propagation by root cuttings:

Aesculus parviflora (Bottlebrush Buckeye)
Albizzia julibrissin (Silk-tree)
Calycanthus (Sweet-shrub)
Campsis (Trumpet-creeper)
Catalpa (Indian-bean)
Celastrus (Bittersweet)
Cladrastis (Yellow-wood)
Clerodendrum trichotomum (Glory Bower)
Halesia (Silver Bell)
Hypericum (St.-Johns-Wort)
Ilex (Holly)
Koelreuteria (Goldenrain-tree)
Maclura (Osage-orange)
Paulownia (Empress Tree)
Rhus (Sumac, Smoke-tree)
Robinia (Rose-acacia, Clammy Locust)
Sassafras
Stephanandra
Wisteria

Some writers claim that lilac can be propagated by root cuttings. I was successful in rooting pieces of the underground parts of lilac suckers, but pieces of true root always failed.

Herbaceous plants that can be propagated by root cuttings include:

Acanthus mollis (Bear's-breech)
Anchusa (Alkanet)
Anemone hupehensis japonica (Japanese Anemone)
Bocconia (Plume-poppy)
Dicentra (Bleeding-heart)
Echinops (Globe-thistle)
Gypsophila (Baby's-breath)*

* *Not the double-flowered kinds when they are grafted on* G. paniculata.

Oenothera (Evening-primrose)
Papaver orientale (Oriental Poppy)
Phlox paniculata (Garden Phlox)
Primula (Primrose)
Stokesia (Stokes-aster)
Trollius (Globe-flower)
Verbascum (Mullein)

Although the plants originated from root cuttings normally produce plants exactly the same as the parent, there are a few exceptions, notably in the case of *Bouvardia* variety 'Bridesmaid', and some of the fancy-leaved pelargoniums (house geranium).

Digging for roots of Japanese anemone. A deer chewed up most of the parent plant in the background.

The roots are laid out on a patch of sand so that they are more readily seen.

They are cut up into pieces about 1 to 2 inches long and placed in a flat prepared as suggested in the introduction.

Kept in a cold frame over winter, they look like this the following May. One plant in the foreground, left, has been dug up to show the new roots.

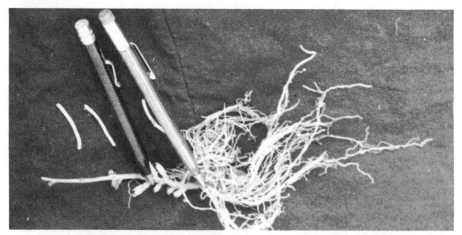

A one-year-old phlox plant, which grew from a piece of root similar to those shown left of the black pencil. The piece of root from which this plant grew is contained between the points of the two pencils.

Cuttings of herbaceous perennials

These can be made at various seasons, depending on the kinds. For example, chrysanthemum and hardy aster, which bloom late in the year, can be propagated by cuttings made in early spring, when the young shoots are starting out from the rootstock. Delphinium cuttings may be taken in the spring and also in the summer when new shoots start from the ground line. With all the above it is desirable to make the cuts just below the surface when the shoots are from 3 to 6 inches long. When chrysanthemums are required in quantity, the tips of rooted cuttings may be used as cuttings when they are 8 to 10 inches tall by removing the top third.

Gypsophila (baby's-breath) can be rooted if shoots are taken in early spring before the flower buds are formed. Viola cuttings can be made in July or August, basket of gold (*Aurinia* [syn. *Alyssum*] *saxatilis*), arabis, and dianthus whenever the new growth is at the right stage—usually just after the flowers have faded.

Chrysanthemum cuttings in foreground taken from a stock plant, February 2, are being inserted in a 1-1-1 mixture, surfaced with sand. Notice the method of making the rooting medium firm at the base of the cutting.

Cuttings on March 15 are well rooted and ready to be potted.

A slightly richer soil mixture, which can be made by omitting the sand from the 1-1-1 mixture, is desirable when they are potted individually in 3-inch pots.

More information on cuttings

Now for a few odds and ends: Wounding the part of the cutting that is to be below ground in many cases is effective in helping to root difficult woody species such as rhododendrons. The wounds can be made by cutting a vertical slit with a sharp knife or by cutting off a slice of bark and wood.

Growth regulators or root-inducing hormones have a place in plant propagation. Though they cannot replace the regular care of the cuttings, they usually can be relied on to induce the cuttings to root earlier and to produce roots in greater numbers. They are obtainable under various trade names. Those containing indolebutyric acid seem to be preferred. No matter which kind you use, be sure to follow the manufacturer's directions as given on the label.

Abrasives, because of their irritant effect, are used sometimes as rooting media. Powdered pumice and powdered glass are said to be used effectively in rooting difficult alpines. The pumice is mixed in the proportion of five parts to two parts sand; powdered glass, three parts to one of sand. You can probably get the pumice from a hardware store—ask for their coarsest grade; the glass might be obtained from a sandpaper manufacturer, or you can make your own by putting glass in a bottomless box on a concrete floor and pounding it with a road rammer.

Those who are interested in the kinks and quirks attendant on the rooting of cuttings should look over the Propagation Notebook in the Appendix. The cuttings were inserted in a cold frame as described above.

Using polyethylene

The plastic film, polyethylene, has become an invaluable aid in rooting cuttings and as a covering for the rooting medium in air-layering. This material has the faculty of permitting the passage of air and to a large extent preventing transmission of water vapor. Polyethylene is available in various thicknesses ranging from $\frac{1}{1000}$ of an inch to $\frac{8}{1000}$. According to Donald Wyman in the *Arnold Arboretum Garden Book*, the amount of water vapor transmitted during twenty-four hours ranges from 0.95 to 0.15. This represents the number of grams lost per 100 square inches of exposed film surface during twenty-four hours. The method is to insert the cuttings at the right time for each species. Then the cuttings are covered with the plastic, put on in such a way that there are no openings of any kind. The following pictures show a technique that can be readily adapted by home gardeners.

Probably my poor results can be attributed to the fact that cuttings of some species were not inserted early enough in the season and the covering was not tight enough to retain moisture.

A flat is filled with a mixture of sand and peat moss—75 per cent sand, 25 per cent peat moss by bulk. A tamper, in this case consisting of a section of flooring with two screw hooks, which serve as a handle, is used to firm the rooting medium. The left half has been pressed down so that the surface is ½ inch below the rim of the flat.

It is important to prevent wilting of cuttings. To this end the work is done in the shade and, whenever a row is finished, it is watered in thoroughly.

The method of inserting cuttings with the aid of a dibble (dowel section or stick). A hole is made, the cutting put in with its base touching the bottom of the hole, and the rooting medium is firmed by pushing the soil around the base with a dibble as shown.

Another method is to make a slit in the rooting medium. The tamper serves as a guide for the kitchen knife which is used to make the slit. The cuttings are inserted and the soil is firmed from the side.

"The covered wagon!" The flat has been filled with cuttings and plastic is applied and tacked down. The half hoops can be made from wire scraps on hand —or cut-up coat hangers.

In this picture representative samples of each kind of cutting were removed on September 24. Only one (*Cytisus* x *praecox*— early broom) had formed roots.

So the plastic was reapplied and the flat was put in a cold frame with the hope that the unrooted cuttings would come through the winter and perhaps start a new growth the following year.

As soon as the cuttings were placed, vermiculite was poured in around to help avoid violent fluctuations of temperature. It will be noticed that the potted plants, in foreground and at right, also were covered.

The flat of cuttings as it appeared the following spring. The results were disappointing. The cytisus had rooted 100 per cent; the remainder varied, some being dead and the others in no case had more than about 10 per cent rooted.

It is interesting that *Cytisus* x *praecox* is the only species that rooted 100 per cent. This surprised me because I had looked on this as being difficult to root, especially when two-year-old shoots were used. In my previous experience with this plant it was the current season's shoots only which were successful.

Polyethylene is a versatile material. In addition to the use cited above it can serve as a cover for grafted scions in top-worked trees; and, cut into narrow strips, it can be used to tie in buds.

Young rhododendron plants in a polyethylene-enclosed shelter. The material is used by home gardeners and commercial growers alike in many phases of plant propagation.

The plastic "covered wagon" method for rooting cuttings worked well in this case. Cuttings of heaths (*Erica*) and heathers (*Calluna*) were put in the flat in early summer. The plastic was placed over the wire frame and the flat was left outdoors until the following spring. Then the rooted cuttings were set in the open ground.

Constant mist

In recent years considerable interest has developed in what is called the "constant mist" method for rooting cuttings. Essentially the method is one of rooting cuttings in the open in full sun, preventing them from wilting by keeping the foliage constantly moist by means of nozzles which deliver water in a fine mist. Actually as the system has evolved, it is used mostly in glass or polyethylene houses and the term "constant" is really a misnomer since most commercial setups now use intermittent applications of mist that are set to function automatically. Fog propagation is another variation, differing from the mist method mainly in the automatic dispersal of water as fog. The equipment for fog propagation is considerably more expensive.

Although mist and fog propagation are methods of prime importance to commercial growers, anyone interested in propagation is intrigued by them. For my first mist setup, I contrived an enclosure of polyethylene attached to builder's laths. The reason for making this corral was that a windy spot was the only convenient place to install it; and it seemed to me that it would help to confine the fog and prevent it from being blown away from the plants. Even though it is unnecessary to keep the mist going during the hours of darkness, the mist was left on day and night from early June until frost. The second year the corral was dispensed with and a smaller nozzle was installed, which seemed to work as well or even better than the enclosed system. Apparently there was enough variation in the direction and force of the air currents so that almost invariably no part of the area, ordinarily covered by the spray, was without moisture long enough to injure the cuttings.

Sven Sward, who was horticulturist at Vassar College, after he saw my first setup, devised a similar but smaller box which was installed in a greenhouse. He had no difficulty in rooting a variety of plants. (See page 145.)

Guy wires at each corner of the enclosure were fastened to heavy stakes to anchor it.

This shows the enclosure which was devised to serve as a windbreak for a "fog box." It was made by tacking polyethylene onto 4-foot building laths. Two sides were removable. The purpose of the grating, of pine 1 by 1 inch, was to eliminate any danger of the rooting material becoming waterlogged. The fog came from the nozzle attached to a piece of flexible copper tubing, which in turn was attached to a hose connected with the domestic water supply.

The boxes were about 1 foot square and 6 inches deep; 4 inches would be better because half a cubic foot of sand is too heavy for convenient handling. A strip of metal fly screening is put over crack in the box to prevent the rooting medium from sifting through.

A 2-inch layer of coarse sand is put in to ensure adequate drainage.

Two broom clamps were used to hold the removable side in place.

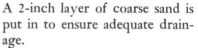

Humidification without windbreak. The "arrow" points to the nozzle. This was a Bete fog nozzle, C. series, which gave perfect performance in making a very fine mist without any clogging. It was a great improvement over the preceding year, when nozzles of a different make were constantly clogging.

A similar fog box on a smaller scale set up at Vassar College under greenhouse conditions.

Commercial use of the constant-mist system in a deep cold frame.

Another commercial use of the mist system in a greenhouse.

A more modest misting system that can be contrived by a home gardener. The structure is 6 feet by 6 feet and as shown here is only partially enclosed by plastic, but the bent pipe supports can easily be totally covered, making it a miniature greenhouse suitable for several kinds of propagation.

Cuttings of *Daphne cneorum* (rose daphne) from vermiculite —50 per cent rooted.

From peat and sand—90 per cent rooted.

From sand—100 per cent rooted. It is interesting to notice that the larger cuttings containing two-year-old wood as well as that of the current season rooted as well as, or better than, the new wood. All of the above were inserted on June 19; they were rooted by the middle of July.

At the end of July it was decided to move these plants from the fog box. Accordingly a flat was prepared by putting an inch or so of partly decayed leaves in the bottom, which was surfaced with 2½ inches of sifted 1-1-1 mixture. (See Chapter 10.)

This picture shows a technique of planting rooted cuttings. The soil is not at first packed down, which allows the holes to be made easily with the fingers, to receive the roots. Firming is done after the roots are in place.

One of the advantages of the constant mist is that cuttings larger than would normally be used often root with greater ease than the orthodox smaller cuttings. Here are cuttings of Meyer's lilac (*Syringa meyeri*). Note the good root system developed on three of the larger cuttings.

Two of these were potted immediately. They were returned to the fog box for a week or two, then plunged into soil in partial shade.

One interesting thing about this system of plant propagation is that some cuttings make an excessively large callus. Varieties even within the species are more prone to this than others. A plant propagator does not like to see excessively large calluses because apparently roots are seldom formed when these conditions occur. Among the treatments recommended are: dipping the callus in a rooting powder (left) and paring off excess callus growth (right).

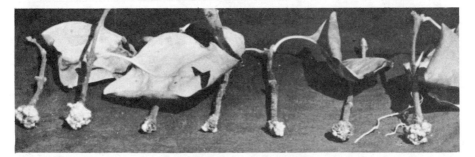

Here are the results. The two cuttings on the left show still more callus growth as a result of the powder treatment. Three in the middle are those that were pared and those at the right, one of which produced roots, received no treatment.

Cuttings of *Rhododendron obtusum kaempferi*, a popular evergreen azalea.

Cuttings of *Cytisus* x *praecox*. Notice that it is the largest cutting that has the most roots.

After two or three frosts the rooted cuttings, including those which had been potted up, were packed as closely as possible in the corral.

Salt-marsh hay was packed between the walls and the flats. A
thin covering was put over the
plants.

The two removable sides were
loosened at the top and brought
together to make a sort of
peaked roof.

The results from cuttings subjected to constant mist were astounding. Large cuttings, in general, rooted better than the small
cuttings, of current-year wood, which used to be considered essential. Plants that were thought to be somewhat difficult—rose
daphne, cytisus, and the star magnolia—rooted with ease. Cuttings
of purple lilac rooted 100 per cent, while those of a white lilac
(shown on page 149) were less successful—only one of seven
rooted.

Sven Sward, in his mist experiments, was successful in carrying over winter in a cool greenhouse the following plants: American holly (*Ilex opaca*), dogwood (pink-flowering), false-cypress (*Chamaecyparis*), leucothoe, *Picea abies* 'Nana', *Metasequoia glyptostroboides*, *Pieris japonica*, rhododendron, *Viburnum sieboldii*. These were potted in September and planted out in a cold frame the following March. During the following winter they were protected by a covering of leaves and glass sash. Mr. Sward reports that about 70 per cent of the plants rooted; rather surprisingly, hemlock was a complete failure, probably a fluke.

Early broom (*Cytisus* x *praecox*) and lavender are xerophytes —that is, plants that are adapted to growing under very dry conditions. Surprisingly, they reveled in the humidity. One might expect that fungus diseases would be rife under mist, but such is not the case. However, commercial growers today often rely on fungicides, such as captan, or make a point of frequently ventilating plastic-enclosed propagating houses to control diseases.

Rhododendrons, once considered slow and often difficult to propagate from cuttings, root very well under mist. Most growers take the cuttings in early summer, then keep them over winter in cool greenhouses at temperatures of around 40° F. Then the temperature is raised in early spring to as high as 70° F. during the day. The rooted cuttings respond to this treatment with a burst of growth, resulting in larger, salable plants in record time.

4 BULBS, CORMS, TUBERS, RHIZOMES, ROOTSTOCKS

Catalogues and gardening books often lump all these together under "bulbs," which makes some persnickety botanists gnash their teeth. This is how they differ:

Bulbs are made up of closely packed, fleshy scale leaves attached to a comparatively small basal plate of solid tissue; examples are tulip, daffodil or narcissus, and lily.

Corms are similar in appearance to bulbs, differing in being solid throughout with small buds ultimately forming on the top; examples are gladiolus and crocus.

Corm of gladiolus at left; bulb of daffodil at right.

Cutaway view of tulip bulb, showing tiny blossom and closely packed leaves attached to basal plate.

Tubers may be swollen roots, as in the dahlia, tuberous begonia, and cyclamen, or swollen stems, as in the case of the Jerusalem-artichoke and the potato.

A *rhizome* is a thickened shoot that grows underground or along the soil surface, producing roots from its undersides and leaves or shoots above the ground; examples are the bearded iris and Solomon's-seal. When these shoots are short and stubby, as in canna, they are named *rootstock*. Trillium, according to *Gray's Manual*, is a "praemorse tuber-like rootstock."

Bulbs, corms, tubers, rhizomes, and rootstocks are storage organs full of food which enable the species to survive unfavorable climates. *Bulbils* are the little bulbs that form on the aerial stems of lilies and other plants; *bulblets* are those that form underground.

These specialized structures in most cases afford us ready means of propagation (the exception being cyclamen), because at some time in their growth cycles they are more nearly dormant than plants without these structures and can be manipulated without much injury.

Bulbs

Many bulbs propagate themselves naturally by breaking up into a number of small bulbs. Tulips do this especially when they are planted shallowly. In the case of daffodils the mother bulb makes offsets which are known as "slabs."

The hyacinth is a shy propagator under natural conditions, but by processing a dormant bulb it is possible to increase greatly the numbers of bulblets. There are two methods commonly used— "scooping" and "scoring." The first named results in a greater production of bulblets, but usually they are slightly smaller than those produced by scoring. A potato- or melon-baller is an ideal tool for scooping. If you do not have ready access to such a tool, you may file an unmatched stainless-steel teaspoon to make a sharp edge and use it instead.

This picture shows the beginning of the removal of the basal plate to expose the scale leaves which make up a large portion of the bulb.

The job is completed.

In scoring, three cuts are made in the base, extending about a third of the way into the bulb.

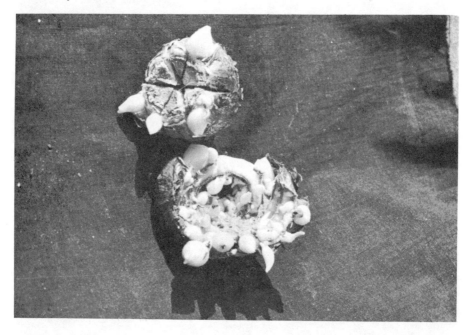

The bulbs are placed upside down in a temperature of about 70° or 80° F. After a lapse of two or three months they looked like these, which should be planted in the fall, still upside down, covered with two or three inches of soil.

The following year they are dug up, when the foliage has withered, and separated. They are then planted about 3 inches apart in rows 1 foot apart, to grow into bulbs of flowering size, which may take from three to six years.

When lilies have been growing in one spot for a number of years, the bulbs are likely to become too crowded for best results. The ideal time to divide them is when the foliage is becoming yellow.

Here is a "double" bulb of the madonna lily being pulled apart. Do not cut off the roots.

The bulbs are reset from 6 to 8 inches apart. Madonna lilies are planted so that there are no more than 2 inches of soil covering them. Most lilies, however, require deeper planting than this.

Many kinds of lilies propagate by means of stem bulbils—*Lilium sulphureum*, for example. There are at least two ways by which the number and size of bulbils can be increased. One is by removal of the flowers in the bud stage, as was done with the one at the left.

Another method is to pull up the lily stems and heel them in as soon as the flowers show signs of wilting.

This uprooting (an injury to the plant) may result in numerous bulblets such as with these (*Lilium hollandicum*).

Madonna lily (*Lilium candidum*) at left; *Lilium dauricum* at right. The two growing points on *Lilium candidum* indicate that the following year there will be at least two flower stalks from this bulb. Notice the bulblets on the underground part of the stem of *Lilium dauricum*.

The plant of *Lilium dauricum* denuded of the bulblets which have been planted separately. Also a dozen or more of the outer scale leaves have been removed for propagation. About one half of them were completely buried ½ inch deep in the flat. Those remaining were planted with the tips showing above the surface.

Two months later.

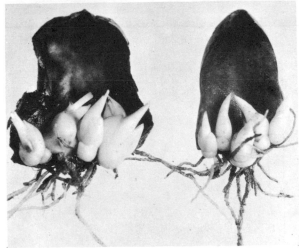

Bulblets produced from scales of *Lilium leucanthum.*

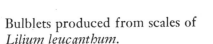

Corms

Recently dug specimens of gladiolus, showing new corms produced on top of the old ones that were planted in May.

Four new corms plus four cormels were produced from the corm in the center of the cluster. The corms are of flowering size. The cormels will take a year at least before reaching flowering size. Ordinarily many more cormels are produced. These can be planted in the spring in rows in the same way as you would plant peas.

Tubers

Tuber of gloriosa-lily (*Gloriosa rothschildiana*). The growing tip is at the left.

Planting in a 6-inch pot. This one had to be put in diagonally because it was too big to be placed upright.

The increase by the end of the growing season. The old tuber is in the middle with two new ones, right and left. Division of the tubers is another way of increasing one's stock of gloriosa-lily.

Tuberous begonias can be multiplied by dividing the tubers. It is necessary or at least desirable to wait until new growth has started before any cutting is done.

Knife blade shows the method of making the cut when two shoots are growing close together.

It is advisable to treat the cut surface with a fungicide. Here a mixture of 10 per cent fermate and 90 per cent dusting sulphur is being used. The applicator consists of a tuft of lamb's wool fastened to the end of a straightened-out paper clip, but a paintbrush would also do.

Tuberous begonias can also be propagated from stem cuttings.

These can be rooted in a standard 1-1-1 mixture with which a 3- or 4-inch pot is filled without packing it down. The hole to receive the cutting can be made by pushing the finger in the center. The cutting is then put in with its base resting on the bottom of the hole and the soil is firmed by pressing it down with thumbs or fingers.

The leaf surface can be reduced by cutting off a portion of one of the leaves.

The cutting is then put in a plastic bag (see page 85) to maintain moist atmosphere, which prevents it from wilting. It is advisable to keep the cutting in the shade until it is well rooted.

Inserted during July, by September the cutting (right) had rooted and made enough new growth to blossom.

Dahlias can be propagated by division of the rootstock. It is essential to have a portion of the stem attached to the root.

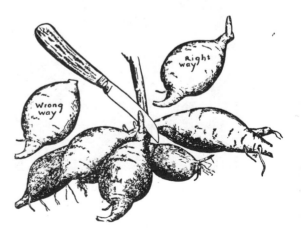

Tuberous-rooted begonias can be propagated from seeds, which must be sown indoors in January or February to be sure of getting flowers from them the same year. Follow the procedure for small seeds (see pages 23–29). Named varieties or any specially good ones you may have raised from seeds can be propagated by cutting the tubers and by making stem cuttings as described above.

Rhizomes

The most familiar of rhizomatous plants are the tall bearded irises. They can be increased in short order by dividing the rhizomes between July and October.

A clump is dug up and the old flower stalks are cut off close to the rhizome.

The leaves are cut back to within a third of their length. They are now ready for planting in a hole with a ridge of soil on which the rhizome is placed, with the roots spread out on either side. (This division is capable of being further divided to make two plants.)

The roots have been covered with earth which is firmed by standing on it.

Rootstocks

Among the native plants that might be considered difficult to propagate either sexually by seeds or asexually by division of the rootstock is the trillium. The pictures which follow show how, by surgery, it is possible to hurry up nature.

Washing away the soil of a potted plant to expose the crown of the rootstock, without disturbing the roots, reveals the growing bud of next year's plant already well along in July. A notch is cut just below the bud at the point indicated by the knife. The plant is then recovered with soil and the pot plunged in a shady spot.

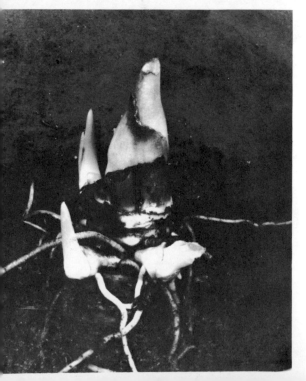

In October, three months after the notching, results are plain. The large bud is the one shown above. Three strong new shoots have broken out just below the notch, and there is evidence of more coming. The only reason for uncovering the plant at this stage was to get this picture.

The first spring after the surgical treatment seven plants appeared instead of the two whose stems are shown in illustration. Four of them bore large flowers, although only two are visible here. The pot was posed to enable you to count the stems rather than to display the blossoms.

In July of the second year our busy clump, uncovered just to satisfy curiosity, showed several little new rootstocks and many more buds.

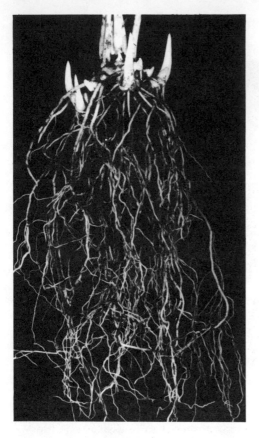

In October the whole clump was removed from the pot for separation. Five new rootstocks were strong enough to start alone.

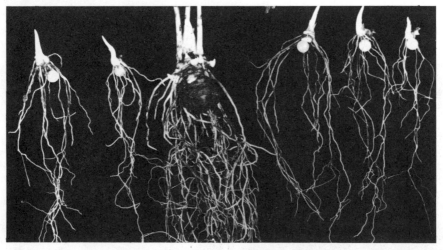

The five new rootstocks have been cut from the mother plant and are ready to be potted up alone. (The white disc under each is a pinhead.) Original rootstock continues to make new offsets.

The second spring the five new rootstocks produced eight. The original rootstock produced six, of which four have flowers despite division made the previous fall.

The photograph shows dividends paid in less than three years. The new rootstocks produced nine plants, eight of them with blossoms. The original rootstock increased to sixteen and now is ready for further division.

And here is how *not* to propagate a trillium. Complete decapitation of the rootstock proved to be fatal.

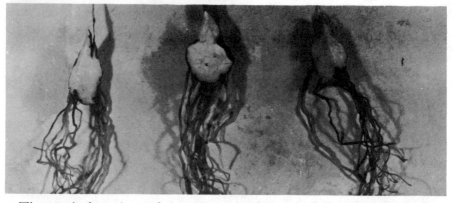

The vertical cuttings of the rootstock, although they had strong-looking roots, failed to grow.

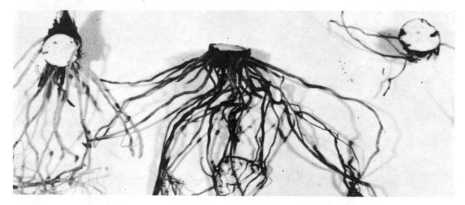

Horizontal cuttings of the rootstock also rotted away. The evidence of these unsuccessful methods of cutting, while perhaps not conclusive, suggests that they be avoided. The operations might have been successes, but the patients died.

5 RUNNERS, OFFSETS, STOLONS, AND SUCKERS

The specialized shoots named above provide easy means of propagating the plants that bear them.

Runners are slender shoots which start, usually, near the base of the mother plant and produce roots and shoots at intervals. A well-known example is the strawberry. Sometimes the flower stalks serve a similar purpose, as in the case of *Chlorophytum comosum*, sometimes called spider-plant or airplane-plant, and *Neomarica northiana* (apostle-plant).

An *offset* is similar but with a shorter stem; an example is *Sempervivum* (houseleek or hen-and-chickens).

A *stolon* is a shoot that starts above the surface of the ground and then bends over so that it comes in contact with the soil and forms roots. An example is *Cornus sericea* (red-osier dogwood). In some plants roots are formed only at the tip of the shoot, as in the case of black raspberry.

A *sucker* is a shoot that starts below the ground from a root or underground stem. Examples are lilac and red raspberry.

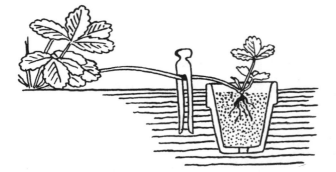

Runners

Strawberry is commonly propagated by means of runners, which in most varieties are formed in great abundance during the growing season. Under home garden conditions these runners can be rooted in 3- or 4-inch pots, filled with rich soil, by pegging them down with a clothespin or something similar. The big advantage of this is that they can be planted without any setback in August, enabling them to produce a crop the following year.

Another runner-forming plant is the creeping bugle (*Ajuga reptans*). The small plants can be dug up at almost any season for increase, or they can be left in place to form a matted ground cover.

Some of the bromeliads (relatives of the pineapple) can be propagated by means of runners such as the one seen here, *Bromelia lasiantha*. These can be handled by removing the runner from the parent plant and putting it in as a cutting or can be rooted the same as strawberry, by putting a soil-filled pot near the plant and waiting until it has formed roots before taking it away from the mother.

This is *Saxifraga stolonifera*, a plant often called strawberry-geranium or strawberry-begonia. These common names are misnomers, because the plant is not strawberry, geranium, or begonia. If you must have an English name for it, call it mother-of-thousands. It is one of the easiest of all plants to propagate by taking the plantlets from the runners and putting them in separate flower pots —3-inch size is ample to start them. (See also page 53.)

Spider-plant or airplane-plant (*Chlorophytum comosum*) achieves same effect by means of its flowering stems, which, when the flowers have faded, produce numerous plantlets.

Offsets

Cobweb houseleek (*Sempervivum arachnoideum*) is an example of a plant which makes offsets. These can be pulled off at any time except during winter, and inserted where they are to grow. The plants to the left, *Saxifraga paniculata* (syn. *aizoon*) varieties (silver saxifrage), with white-margined leaves, have offsets with even shorter stems. These need more care in handling than the houseleeks. The entire plant should be dug up during the summer, the offsets pulled off and planted in individual 2½-inch pots, using a 1-1-1 soil mixture plus an added part of chicken grits or sharp sand. They should be kept in a cold frame, shaded and watered, with ample ventilation at all times until they are rooted.

Stolons

Black raspberries propagate themselves by means of stolons, which in this picture may be seen arching over from the main plant in the background.

This is a closer view of the above.

Later the stolon develops a large root system suitable for planting.

Suckers

Suckers: Red raspberries ordinarily are propagated by digging up suckers which appear alongside the main row. They can be dug up in the spring, cut back about one half, and transplanted.

6 AIR-LAYERING AND LAYERING

This is a means whereby plants which are difficult to root from cuttings can be propagated. There are three methods whereby this can be done—ground-, mound-, and air-layering. In ground-layering it is necessary that the plant to be propagated should have branches so disposed that they can be brought down to the ground and held there until sufficient roots are formed on the layers to permit them to be severed from the parent plants, for example as in the case of the rhododendron pictured on pages 188–89.

Mound-layering is a commercial method used considerably in Europe, but in most cases it is only of academic interest to the amateur. Briefly, it involves cutting back a parent plant almost to the ground level in the spring. This causes abundant shoots to be produced suitable for this type of propagation. The following spring the parent plant is covered with a mound of earth. After a year or so the shoots make sufficient roots to enable them to be dug up and planted.

In air-layering a suitable shoot is selected, the stem is wounded by removal of a cylinder of bark, by cutting a notch, or by making a slit in the stem at the point where it is desired that roots should form. The old-fashioned way, suitable when done indoors (see drawing on page 182), was to wrap the wound with a handful of sphagnum moss and cover it with waxed paper, tin foil, or some such waterproof covering, leaving the top open so that water can be applied to keep the moss moist. Today, however, with the availability of polyethylene, the moss is wetted to the right degree, put in place, covered tightly with the sheet of plastic, which has the faculty of keeping moisture in so that no watering is required until the rooted layer is severed from the parent plant. Any failure you might encounter with this method can probably be attributed to the rooting medium's becoming too wet, due to rain seeping into the ends of the plastic which may not have been tightly closed. To avoid this, use a waterproof tape to seal the ends of the plastic. The best time to air-layer hardy plants is in the early spring. The Arnold Arboretum experimented with a large number of different plants, some of which rooted 100 per cent while some failed completely. See the list in the Appendix.

Another hurdle which has to be overcome is that of caring for the rooted layers in such a way that they survive. This may involve potting the rooted layer and keeping it for a while in a closed cold frame or plastic-enclosed structure and gradually inuring it to out-door conditions by progressively admitting more air.

Air-layering indoors

This dracaena with its 3-foot-long leafless stem was obviously overdue for an operation. Arrow indicates point of wedge-shaped cut, half through stem.

Wet sphagnum moss is wrapped around stem over cut and enclosed in plastic.

Two weeks later, plastic and moss are removed and new roots are visible.

Rejuvenated plant is cut from original stem.

The result after potting—a much more presentable plant.

Air-layering outdoors

Sometime in June a "tongue" is cut in the rhododendron shoot to be layered and is dusted with root-inducing powder. Under the tongue a little moist sphagnum moss is stuffed.

The whole is swaddled with a fist-size ball of moss.

It is then wrapped with sheet
plastic, being careful to tie it in
such a way that moisture cannot
enter.

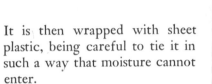

About October 1 the plastic is
removed and if roots show good
development, the moss is partly
removed, with care given to
avoid injury to the roots. The
layer is cut off just below the
root ball and the new plant is set
in a cold frame to become established before winter sets in.

Ground-layering

Soil is improved for ground-layering a rhododendron by mixing sand and peat moss in existing soil.

The branch selected is shown held in the left hand of the operator.

A slit is cut in the stem at the point where roots are wanted.

The "tongue," made by cutting into the stem, is held open by a toothpick.

The wound is dusted with a root-inducing substance applied with a soft brush.

A trench 2 inches deep has been made and the shoot is held down by one hand. The other hand is being used to cover it with soil.

A sizable rock is then put over the layer and serves a double purpose of keeping the layer in place and of helping to conserve the moisture in the soil.

A year later the layer is ready to be transplanted. Notice the mass of roots, especially at the point where the root-inducing hormone was applied.

This clump of sedum (*Sedum spectabile*) reveals two naturally formed layers, ready to be severed from the main clump and replanted. The remaining clump was propagated by division.

7 PERENNIAL VEGETABLES

Most vegetables are either annuals or biennials which are treated as annuals and raised from seed as described in Chapter 1. The propagation of the most important perennial vegetables is described below.

Asparagus is started from seeds. They are planted in the spring in rows 18 inches apart, the seedlings thinned to 3 inches apart. The following spring they are set in their permanent position, and are spaced 18 inches apart in rows 4 feet apart. Wait two years before cutting. If you are in a hurry for an asparagus harvest, you can buy plants and gain a year, but the plants you grow yourself will be superior to the often dried-up crowns that are sold.

Horse-radish is propagated by root cuttings 6 to 8 inches long—about ¼ inch in diameter. The side roots are taken off when the crop is dug in the fall and kept over winter, buried outdoors in soil. They are planted in the spring. In order to avoid planting them upside down, the top of the cutting is cut off squarely and the lower end is trimmed to an angle of about 45°.

Rhubarb is propagated by division of rootstock in the fall or very early in the spring. See peony, page 70, for techniques.

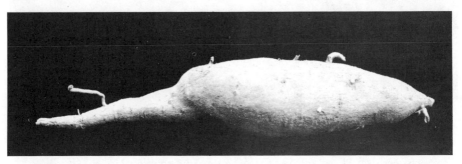

Sweet-potatoes are propagated by means of rooted shoots which are called sets or slips. This is a root tuber from which shoots already are emerging.

Enough slips can be obtained for home use by starting them in a box similar to this one pictured—6 to 8 inches deep, 1 foot wide, and about 14 inches long, filled with moist peat moss and/or sand, or vermiculite.

After a few weeks' time there is a regular forest of shoots. The box containing these should be removed from the house and put in a cold frame or sheltered place for seven to ten days to harden off the shoots before planting them outdoors. When ready to plant in the outdoor garden, they are taken off by holding down the tuber with one hand and pulling off individual shoots with the other.

Do not pot up the slips before planting them. If you do, your crop probably will be like this instead of being similar to the one below.

Normal production of sweet-potatoes. This is a small crop because it was grown so far north (in the mid-Hudson Valley), where the season was not long enough to produce a full crop. Most mail-order seed and nursery companies offer rooted plants that are best for northern gardeners. The variety 'Puerto Rico' is a bush form that uses less space.

White or "Irish" potatoes are propagated by underground stem tubers, either by cutting them into pieces, each one containing an eye, or, as is done in England, by planting the entire potato.

Potato tubers used for planting should be about the size of a hen's egg. Larger tubers may be cut into sections (as shown here), making sure that each piece possesses one–two buds or "eyes." The cut surfaces should be allowed to dry for a few days before planting. It is important that only certified disease-free "seed" potatoes are used. They are offered by most mail-order seed and nursery firms.

Small "seed" potatoes being planted in a trench that is about 5 inches deep. The rows should be about 2–3 feet apart.

8 PROPAGATION OF GRASSES

LAWN-MAKING

Establishing a lawn involves propagation (even though most gardeners do not think of it in this light), since the methods used are seed sowing and various forms of division, such as the cutting up of mature sections of grass into sod, plugs, and sprigs. The preparation of the soil, no matter which method is used, is essentially the same as that described earlier for Sowing Seeds Outdoors (see page 34). However, since lawns are planted for permanence and are often subjected to wear and tear that would ruin other plants, more attention to fertilizer and lime requirements is necessary than for most annual plantings. Most soils can benefit from additions of ground limestone—about 50 to 80 pounds per 1,000 square feet—unless tests indicate otherwise. An application of a complete fertilizer, formulated for grass and used according to directions on the bag, should also be made prior to planting.

Once the lawns are established, fertilizer requirements continue to be high compared to most other plants, and one or more feedings per year are given. Lime applications can be made every five or six years. The kind of grass or mixture of grasses chosen depends on climate and region, exposure (most lawn grasses perform best in full sun but there are some that tolerate partial shade), and use intended for the area. Local outlets, such as hardware stores, nursery and garden centers, usually stock the best mixtures of grass seed or grass divisions (sod, plugs, and sprigs) for their regions.

Seeding

This is the most popular and economical method of establishing turf and is usually carried out in early spring or early autumn in the North. The latter is the more favorable of the two seasons as the young grasses have a better chance of becoming established in the cooler growing conditions that usually prevail at that time of the year. Generally mixtures of several types of grasses are used in the North and include species, varieties, strains, and forms of bent grasses, bluegrasses, fescues, and, where winters are less severe, zoysia. Those used in the South and other mild climates are bahia, carpet, and centipede grasses for sunny positions in soils that have a low fertility, Bermuda grass, which requires a fertile soil, and St. Augustine (applied by sprigging) and Manila grasses (*Zoysia matrella*) for part shade.

Turning over the soil by hand, although slow, nevertheless is a very thorough method of cultivation suitable for small areas.

Mechanical equipment, such as this rotary tiller, is a faster and less arduous means of preparing the soil for lawn or garden.

The steel rake is ideal for final leveling and collecting stones.

Sowing grass seed by hand. It is scattered close to the surface to obtain a uniform covering. Thick sowing is wasteful. The usual rate for most grass seed is about 2 ounces to the square yard.

Planting grass seed with a push-type spreader. After sowing, lightly rake the seeds into the soil.

Sodding

The formation of a lawn by means of sodding or turfing entails the laying of squares or strips of established grass which are usually cut in the following sizes: 1 foot by 1 foot, 2 feet by 1 foot, 3 feet by 1 foot, and 1½ inches thick.

Although expensive, sodding provides the means of establishing a usable grass surface in a short period of time. Another advantage is that it can be carried out almost any time of the year that the ground is not frozen when the newly laid turf can be watered in time of drought. Try to obtain top-quality sod that is fresh, and after purchase put it in position as quickly as possible to prevent deterioration.

Sod being rolled out over a prepared site so that their points alternate (similar to the way bricks are laid).

After the sod has been laid, a top dressing of soil is applied and brushed into the joints.

Newly laid turf being lightly firmed with a homemade wooden "beater."

Plugging

This method involves cutting pieces of turf into 4-inch cubes with as little root disturbance as possible. They are then planted 1 foot apart in soil that has been well prepared. Only grasses that have a creeping habit of growth, such as Bermuda, centipede, and zoysia, should be used for this form of propagation.

Sprigging

A form of vegetative propagation that is used for grasses that spread by means of stolons or horizontal stems. They are cut into 6-inch lengths and broadcast over a prepared bed and covered with a ½-inch layer of sifted soil. Grasses that may be treated in this way include Bermuda, centipede, bent, and St. Augustine.

Lawns may be formed by planting divisions—sprigs or plugs—of certain grasses that have creeping stems.

9 GRAFTAGE

This is the term used to describe the operation of placing or inserting a portion of one plant (bud or scion) upon another plant (the understock) so that they grow together. Budding usually is performed in the summer at a time when the understock is actively growing and the bark separates readily from the wood; grafting usually is done in winter, when the plants are dormant. There are many different reasons why graftage is practiced by horticulturists. Its chief use is in the propagation of plants that, because of their mixed ancestry, do not come true from seeds; plants that do not produce seed; or those that do not readily make roots from cuttings. Most of our fruit trees are in this group. Apples are budded or grafted on seedling apple understocks, on young plants of special stocks propagated by layering, or on pieces or roots of older trees. Peaches are usually budded on seedlings of the "wild" peach. Pears are budded or grafted on pear seedlings or on quince stocks obtained from cuttings or layers; and so on. The understock and scion must be related. For example, you cannot successfully graft an apple on an oak. See the list in the Appendix, page 241.

Graftage may be employed when a plant does not make satisfactory growth on its own roots, which explains why many nurserymen prefer to propagate by budding or grafting certain varieties of roses and azalea. It used to be a common practice among nurserymen to bud or graft named lilacs on plants or cuttings of privet. This practice has been condemned because it renders the lilacs subject to a disease known as "graft blight." It does, however, result in a salable plant in a shorter time, and, if the grafted lilacs are planted deeply so they form their own roots above the understock (some nurserymen cut off the privet understock as soon as the lilac scion has rooted), the blight is prevented. It is preferable to grafting named lilacs on seedling lilacs, which are likely to sucker badly and choke out the scion, and because they are so much alike, it is practically impossible to distinguish the understock from the scion. However, with mist and fog propagation techniques that speed the growth of cuttings, grafting lilacs will soon be obsolete.

Graftage is used sometimes to adapt plants to soil conditions. For example: plums, which thrive best in heavy clay soil, may be budded on peach to enable them to grow in sandy soil. It is also used to help trees to withstand a severe climate, as when apples are grafted on a rootstock of known hardiness.

The important thing to remember in graftage is that the cambium layer of the understock and that of the scion must be in contact. The cambium layer is a cylinder of actively growing cells between the wood and the bark. When the diameter of the understock is larger than that of the scion, or vice versa, it is necessary to put the scion toward one side to ensure that the contact is properly made. See the illustration on page 215.

Budding

If you have a pet apple tree you wish to propagate—maybe it is one whose name has been lost, or one like this one, which is dying—it is possible to give it a new lease of life by budding it on a suitable understock. This can be a seedling apple tree you may be able to find somewhere on your grounds, or you can dig up a root of pencil thickness from an apple on which you can use a whip graft.

Here is a bud stick (right); on the left it is prepared for use by removing the leaves except for ½ inch of the leafstalk left on to serve as a handle.

The bud which is to form your future tree is being cut from the bud stick.

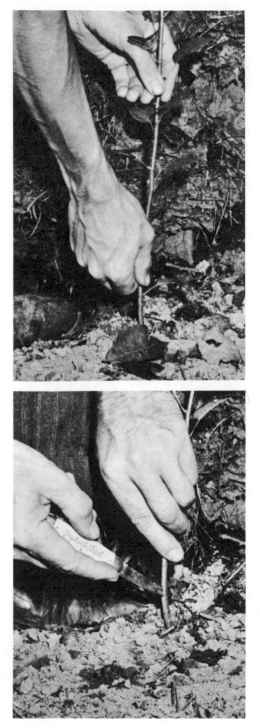

The lower leaves of a seedling understock are removed.

"T" or "shield" budding is the method commonly used. The first step is to make the transverse cut of the "T." The knife must be razor sharp, preferably with a rounded point (see drawing page 206). The cut should be made through the bark but not extending into the wood. A vertical cut is started about 1½ inches below the horizontal cut; as the latter is reached the blade is flicked, left and right, to raise flaps of bark at each corner of the cut.

A bud is slipped into the "T."

The bud inserted and tied in place with a rubber band. The advantage of using rubber as a tie is that it "gives" so that there is less danger of strangling the bud owing to increasing thickness of the understock. It is possible to buy rubber or plastic "ties" especially manufactured for this purpose.

Budding knives.

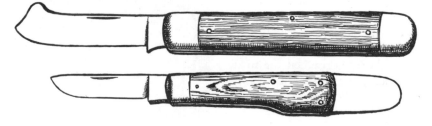

The following spring the bud has started to grow and the understock is cut off just above it.

Seven years later the bud has grown into a tree of flowering size. The only pruning this tree has experienced was done by rabbits the first year and deer the second and third years!

Hybrid tea roses usually are propagated by budding on seedlings or cuttings of a strong growing rose such as *Rosa multiflora*, known for its arching growth habit and multitudes of small, white single flowers in June. The home gardener may have a mature plant or two of *Rosa multiflora* which grew up from the understock (below-ground parts) of a grafted rose, or by chance may have a "living hedge" of this rose which will provide suitable material for budding in the cane. The climbing rose, 'Dr. Huey', also has been used successfully and there may be others. Several buds can be inserted in one cane. These should be spaced far enough apart to permit the cane to be made into cuttings containing at least two or three nodes. The buds can be inserted any time when the bark of the understock lifts (separates) readily. This period usually occurs during July and August or during early spring. See the illustrations on the previous pages for details of budding.

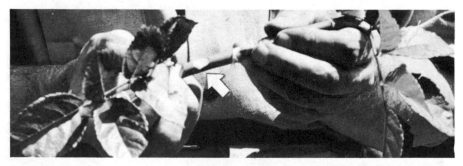

After a week or two the cuttings can be made. In order to prevent the development of suckers it is desirable to cut out the growth bud of that part of the understock which will be below ground.

The bud is tied in place with soft raffia, rubber "budding strips," or, as was done in this case, with a narrow strip of plastic film.

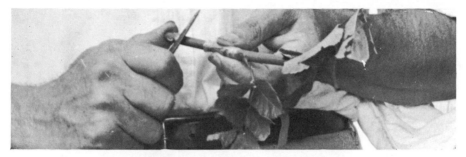

Make the basal cut just below a joint.

The cuttings are inserted in a standard cutting mixture and kept in a closed and shaded cold frame or under constant mist.

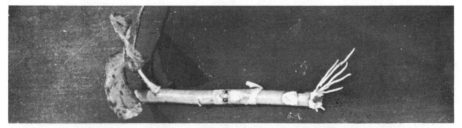

A few weeks later roots have started to grow on the understock.

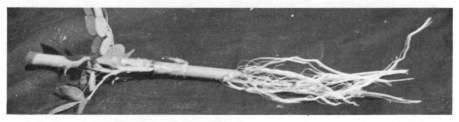

The inserted bud has started to grow.

The following spring the budded portion is blossoming. The under-stock is cut off quite close to the budded portion. This plant is one which was left untransplanted in the cutting frame. Ideally it should have been transplanted in the early spring. Notice the size and quality of the root system. Although it was not planted in the rose bed until June, late planting did not seem to harm it greatly.

One way of propagating lilac is by budding it on cuttings of privet during July and August. Preferably the cuttings should be of pencil thickness.

Grafting

There are many kinds of grafts, including whip, splice, saddle, wedge, side, and veneer. Of these the most important are the whip graft, sometimes called the "tongue" graft, and the side graft, or a modification of this called veneer graft. Then there is the cleft graft, which is used to make over fruit trees, an operation that is known as top-working.

Whip-grafting is commonly practiced in propagation of cherry (both ornamental and fruit-bearing), apple, pear, etc.

The way to make a whip graft is as follows: the scion is cut to a length of 4 to 6 inches from a shoot which grew the preceding year.

With a sharp knife, long sloping cuts are made on both scion and understock. This should be done with a single cut to ensure a smooth surface.

Tongues are made in both by vertical cuts starting near the point of the diagonal cuts.

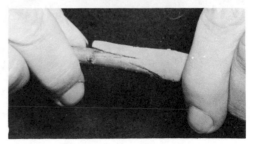

Stock and scion are then fitted together.

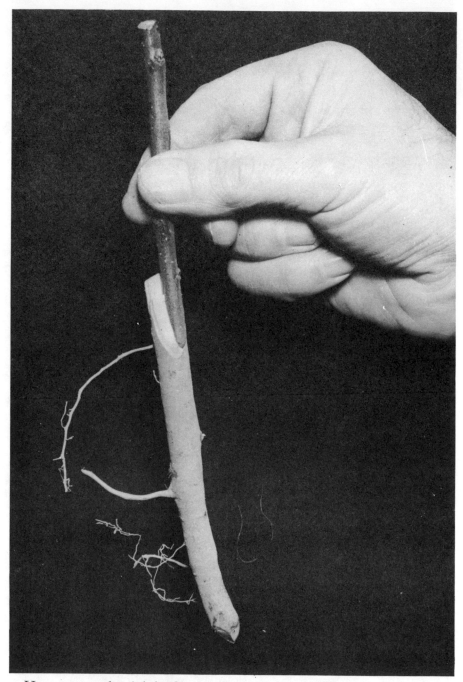

Here we see the tightly fitted joint with the cambium layers of scion and understock aligned at one side.

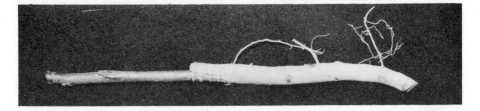

The finished graft tied together with thin cotton twine. Grafts of this kind are stored and treated the same as hardwood cuttings (see page 114).

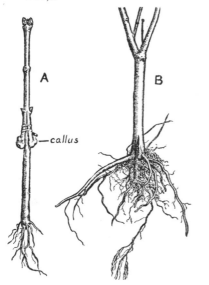

—callus

A saddle graft is made by cutting the understock to a long wedge shape; the scion is split and fitted over this wedge as shown at A. This kind of grafting is considered to be especially desirable in propagation of lilac because the base of the scion is favorably situated to induce rooting. The privet serves mainly as a nurse root as shown at B.

This is a saddle graft of lilac on privet cutting. The base of the cutting and site on the graft have been dusted with a root-inducing powder. It was inserted in sand to the lowermost leaves and put under constant mist, but the union failed to take.

Probable causes of failure are: it was done at the wrong season (we know it can be done during winter on rooted cuttings of privet), and/or the material from which the privet cuttings were made was not fresh—it may have been lying around for a week or more.

This is a side graft often used in propagation of Japanese maples, also for evergreens—both broad and narrow-leaved kinds (conifers). A greenhouse is needed for good results.

The understock usually is a seedling of a related plant which has been potted during the preceding spring so that roots are thoroughly established when the time comes to graft on it during the winter. The understocks are stored in a cool greenhouse (40° to 45° F.) until three or four weeks before the actual grafting is to be done. They are then put in a temperature of 60° to 65° F. to start root action. A slit is cut near the base of the understock; the scion is cut to a long wedge shape and is pushed down into the slit and tied in place. They are kept in a closed propagating case in the greenhouse until the grafts have taken. Then air is gradually admitted. The tops of the understocks are gradually cut back until they are entirely removed.

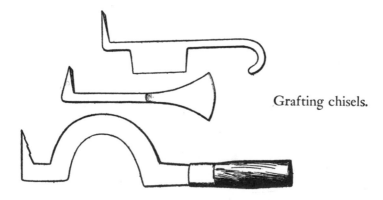

Grafting chisels.

A cleft is made by means of a grafting chisel or a strong pruning knife. A club or wooden mallet is used to drive the chisel into the wood.

The cleft graft is used chiefly to change a fruit tree, such as apple, to a better variety. Branches, preferably between 1 and 2 inches in diameter, are cut off squarely.

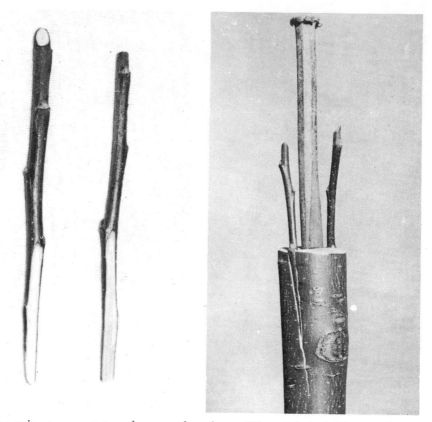

The scions are cut to a long wedge shape. The cleft is held open by the wedge of the chisel, the scions are inserted, the wedge is removed and the "spring" of the wood holds them in place.

All cut surfaces are covered with grafting wax.

This work is done in early spring when the buds of the understock are beginning to swell. It is desirable that the scions should be
in a less advanced stage of growth than the understock. Therefore
they are cut the preceding fall and are buried out of doors in a cold
spot, such as the north side of an unheated building.

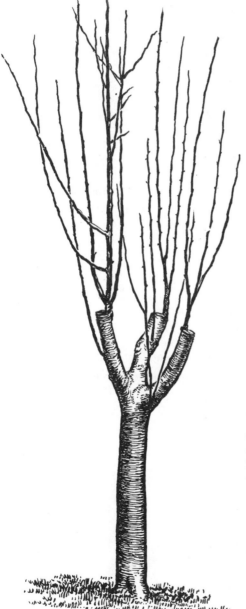

The owner of this apple tree
wanted   to   change   it   to
'McIntosh', so it was cut back to
three branches, in each of which
scions were placed. The shoots
originating on the trunk and
branches below the grafts must
be removed.

Grafting is frequently resorted to by cactus fanciers, either to display their pets more efficiently or because they do not thrive very well on their own roots.

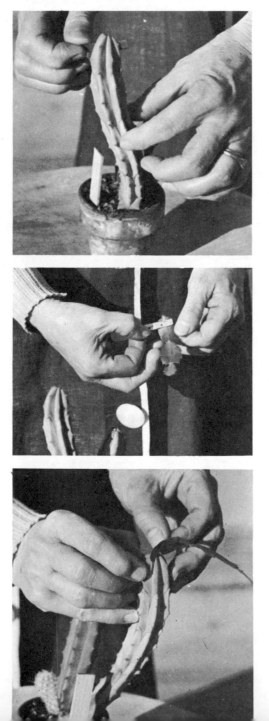

A slit is made in the top of the understock of the blue candle cactus (*Myrtillocactus geometrizans*) by means of a razor blade of the injector type.

The scion of Christmas cactus (*Schlumbergera* x *buckleyi*) is prepared for insertion by cutting a sliver from the base of the stem.

It is inserted in the slit previously made.

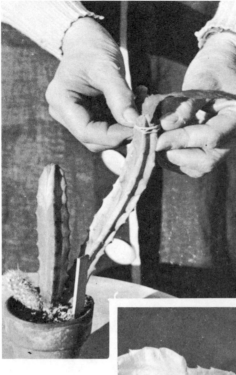

Then it is pinned in place with a cactus thorn. A rubber band is put around the understock to assist further in holding it.

Owing to exigencies of photography the graft did not take, so later the secondary shoot of the understock was grafted with a single joint of Christmas cactus. This one was successful and within a few months had made the growth seen here. The very spiny little cactus on the left is an interloper. It is *Mamillaria elongata*. I had given the parent plant to a Mexican boy who expressed an interest in it, reserving for myself one shoot, which was "temporarily" placed in with the *Myrtillocactus*.

This ensemble resulted from a talk on "Fun with House Plants," which I gave at a garden-club meeting. The top of torch cactus (*Tricho-cereus spachianus*) was cut off and inserted in the rear end of the hol-low, smug-looking cat, where it was intended to grow and form the animal's tail. Then the two scions of Christmas cactus (*Schlumber-gera*) were wedge-grafted in the decapitated torch cactus. The photo-graph was taken about one year later. Since then the *Trichocereus* has provided several tails for similar cats—and even dogs!

Notocactus haselbergii on torch cactus (*Trichocereus*).

A species of *Cephalocereus* grafted on cereus. In this type of graft the top is sliced off the understock and the scion, similarly cut, is put in place and held there by cactus thorns and by tying.

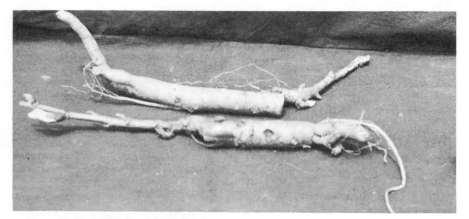

Garden varieties of *Paeonia suffruticosa* (tree peony) are notoriously difficult to propagate by means of cuttings. Because of this, commercial peony growers usually increase named varieties by means of grafting. The understock may be seedlings of *Paeonia suffruticosa;* the wild form of tree peony; or pieces of roots of herbaceous peony. The latter is preferred because it is easier to recognize the understock if it should happen to send up suckers. This shows scions of tree peony which have been grafted on pieces of root of herbaceous peony.

When planting recently grafted peonies such as these it is desirable to set them so that only the upper bud is above the ground.

Propagating bed for camellia varieties in public park at Norfolk, Va. Wide-mouth gallon jars were used to maintain a humid atmosphere around the grafts. The understocks were seedlings of camellia which were cut back almost to the ground preparatory to grafting them (in January in the vicinity of Mobile, Ala., and in February and March around Norfolk, Va.). A wedge graft was used. As soon as the grafting was finished, the jars were put in place, covered with burlap, and the whole bed mulched with a 6-inch layer of leaves. When there was evidence that the grafts had "taken," the burlap was removed and air gradually admitted to the grafts by tilting the jars. However, most camellias are grown directly from cuttings and are on their "own roots."

10 ROOTING MEDIUMS AND EQUIPMENT

Almost any moisture-holding substance can be used as a rooting medium for cuttings. Among those favored by experienced propagators are: sand, peat moss, sphagnum moss, vermiculite, perlite, polystyrene, and mixtures of these in various proportions; then

Vermiculite, perlite, milled sphagnum moss, peat moss are all increasingly used for seeds and cuttings because they are sterile and light in weight. The small fiber flats shown here are a good size for the often limited space under fluorescent lights.

there is the 1-1-1 mixture of garden soil, sand, and peat moss or leaf-mold, an old standby, which has the advantage of containing some plant nutrients and can be used for starting seeds as well as cuttings. Just plain sand works well with a large number of species, that is, if the right kind—coarse, sharp, and porous—is obtained and is watered frequently. The chief disadvantage of sand is its great weight if it is contained in flats which have to be moved occasionally. Sand and peat moss combined, equal parts of each by bulk, are good especially for those plants which require an acid soil. Vermiculite is a lightweight material which is valuable because the roots produced in it are, in the case of yew (*Taxus*) at least, much less brittle than those rooted in sand. Perlite and vermiculite, equal parts of each, have been used with success. The various packaged "soilless" mixes now sold in garden centers and other outlets are simply variations of all these materials—except that they contain no soil. (Of course it is possible to purchase potting mixes that *do* contain soil.) Most of the soilless mixes contain nutrients that will last for a time before the gardener must start applying fertilizers to make up for the lack of soil. All are sterile—an advantage for seed sowing, since the soil-borne damping-off fungi will not be present. Although these packaged soilless mixes are expensive compared to your own preparation of the standard 1-1-1 mixture and other formulations, their convenience for the propagator today cannot be denied. Follow the instructions on the package when using these special mixes. Some plants root with greater ease with one or the other of these materials.

To begin with, I would suggest a mixture of sand and peat moss in proportion of about 75 per cent sand and 25 per cent peat moss, or just sand. For seeds (and some cuttings) it is better to use a mixture of equal parts sand, peat moss, and garden soil (the 1-1-1 mixture). This should be surfaced with a half inch of sand, some of which is pushed down with the dibble (planting stick) into the medium to provide a well-drained and aerated spot at the base of the cutting. If you have the time and are interested, you could amuse yourself by experimenting with other rooting mediums. When soil is contained in the rooting medium, it is desirable to pasteurize it separately, either chemically or by heating it to about 180° F. for half an hour or so in the oven.

One of the best all-around mediums for cuttings is a mixture of sand and peat moss. In the picture peat moss contained in the pail is being moistened preparatory to mixing it with the sand.

Proportions of sand and peat moss vary. Here we have a ratio of about ten parts sand to one part peat moss.

The peat moss and sand must be thoroughly mixed. A good place to do this is in a steel-tray wheelbarrow.

The 1-1-1 mixture of equal parts garden soil, sand, and peat moss is put through a ¼-inch-mesh sieve. The coarse material, which does not pass through the sieve, is known as "rough stuff." It can be used in the bottom of flats or pots to help drainage when rooted cuttings are being "flatted" or potted.

Here is vermiculite being poured into a box which has a 2-inch layer of coarse sand in the bottom to help drainage. One advantage of vermiculite is its light weight.

If you are interested in raising a small number of cuttings of different kinds of plants, flower pots are preferred to a bed in a cold frame or on a greenhouse bench. You see, some kinds root more quickly than others and should be hardened off as soon as they have rooted, which can be done without disturbing the remainder when they are in individual pots. The next best thing is to use fiber, metal, plastic, or wooden trays or flats about 10 by 14 by 4 inches although smaller sizes are available and will serve as well. The last-named should be treated with copper naphthenate (Cuprinol), which helps to prevent decay of the wood.

The use of sphagnum moss for seed germination and for rooting cuttings is a comparatively recent development. Its use practically eliminates danger of seedlings being attacked by damping-off fungi. Sphagnum is a moss (peat moss is decayed sphagnum moss and one of the best sources of humus) which is commonly found growing in bogs. You can buy a quantity sufficient for your needs from garden centers or mail-order garden supply firms. It is available in both "milled" (ready for seed sowing) or "unmilled" form.

If you collect your own moss from a bog or have unmilled sphagnum, it is prepared for seed sowing by drying (if from a bog) and then rubbing it through a sieve, which can be made by tacking to a frame a piece of hardware cloth with three meshes to the inch. It is then moistened and packed evenly in a small flat, watered thoroughly, and an additional thin layer of sphagnum is put on the surface. The seeds are sown on the surface and sprinkled lightly with water applied as a fine spray. The flat is then covered with glass or slipped into a plastic bag. Usually no watering is required for two or three weeks, by which time most of the seeds will have germinated. As soon as this occurs, the covering should be removed and the seedlings gradually inured to ample light in a window or under fluorescent lights.

Shredded moist sphagnum is rounded above the rim of a galvanized asphalt-painted metal flat. (A plastic or wooden flat can be used if more convenient.) A wooden tamper made for this purpose is used to press it down evenly to ½ inch below the rim.

The flat has been prepared for sowing seeds, the moss has been saturated with water, seeds have been sown on the surface of sphagnum and covered with a pane of glass laid on a spacer frame. Instead of the glass a sheet of plastic film could be tacked to the frame.

Seedlings can be kept in a state of suspended animation for a long period. Here are two flats of cinchona seedlings, about twelve months old. Those on the right were taken from the flat on the left two months previously. Sphagnum can also be used for propagating plants vegetatively by stem, root, and leaf cuttings.

Begonia leaf cutting showing new plantlets in sphagnum moss.

Leaf-bud cutting of camellia rooted in sphagnum moss.

Equipment

A sharp knife is necessary for making cuttings and in budding or grafting. A kitchen paring knife, if properly honed, will serve in a pinch. If you are planning to top-work fruit trees, you will need either a grafting chisel or a knife with a heavy blade which won't break when it is hammered into the understock. Special budding and grafting knives are, of course, preferred (see pages 206 and 214). You will need tying materials when budding and grafting is practiced. For budding you can use raffia if it is moistened to make it pliable; or rubber budding strips, which can be made quite easily from rubber bands or purchased; or plastic film cut into strips about ¼ inch wide. If raffia is used, it is necessary to cut through it on the side opposite the bud after two or three weeks. This is to prevent strangulation of the bud by the increase in girth of the understock. For grafting, thin, soft twine which disintegrates in contact with moisture is desirable for tying side and whip grafts in place.

A small cold frame also is desirable. This can be purchased ready-made, or one can easily be made by using scrap lumber like the one pictured on page 99, which was made to fit a discarded window sash. The back should be about 18 inches high and the front 12 inches. The two ends are made by cutting a 6-inch-wide piece of lumber diagonally to give the desired slope. If you don't want to go to the trouble or expense of making a cold frame of this size, you can get by with an old apple box or something similar with the top and bottom knocked out. This should be placed on the ground so that there is a slight slope from back to front, with a cover of glass or of translucent plastic tacked onto the wood frame of lumber 1 by 1 inch cut to fit.

A standard cold frame 3 by 6 feet can be made into a hotbed by installing an electric heating cable with a thermostat. The bottom heat it provides is desirable, but you can get along without it.

Not necessary but a help if you scorn the packaged soilless mixes and prefer to assemble your own soil mixtures are two sieves, one with ¼- or ⅜-inch mesh, the other with a mesh of ⅛ inch. The first can easily be obtained by buying a small portion of hardware cloth. Cut it to the exact size you want and tack or staple to a frame made of lumber 1 by 4 inches. The over-all dimensions

A propagating frame with the sash removed. On the left are *Campanula medium* (Canterbury bells) and *Digitalis purpurea* (foxglove), which were sown at the end of June and transplanted to the cold frame in July. They were photographed late in September. These can be planted in their flowering positions in the fall or left as they are until spring, depending on the severity of the climate. At the right there are various late-sown, recently transplanted perennials which are not big enough to be put in their permanent locations.

should be about 12 by 15 by 4 inches. The small-mesh sieve can be made from wire fly screening. The frame for this can be smaller than the preceding one, say about 6 by 9 by 3 inches.

A "float" or tamper is needed to compress soil or sphagnum moss in the flats. This can be about 6 by 4 by 1 inches made with a handle. You will notice on page 136 one made from a section of flooring with two screw hooks put in to serve as a handle. For tamping soil in pots a smooth-bottomed glass tumbler or a section of a tree branch about 2 inches in diameter can be used.

An example of a ready-made cold frame that can be bought from mail-order garden supply firms. This one is unique in that it is self-ventilating—the plastic cover is opened at 72°F., closed at 68°F. by a sun-powered thermal device.

Homemade cold frame with plastic cover that can be put together with scraps of wood.

Fluorescent lighting

So far the equipment costs have been negligible. But if you want some of the refinements, such as fluorescent lighting, you will have to dig more deeply into your pocketbook. The amount depends on whether you buy one of the fancy, more furniture-like pieces or a scientifically designed propagating case (see page 237), which can be very expensive, or whether you are willing to settle for one of the simple table-top fixtures or something similar to the experimental setup I devised.

Before you decide, you should be reminded that artificial light does not possess the intensity of sunshine, but if the lights are kept on long enough, and if basic requirements of plants as to temperature and humidity are heeded, the results can be excellent. The length of time to which plants should be exposed to artificial light varies, but the majority of plants require about sixteen hours of light and eight hours of darkness in a twenty-four-hour period.

Most cuttings should be positioned about 1 foot beneath the tubes. Cacti and succulents, however, should be about 6 inches beneath the light. In the case of seeds, place the containers at either end of the tubes where there is less light. After germination has taken place, move the containers closer to the center of the fixture and about 4 inches beneath the tubes. All seedlings, such as tomato, pepper, snapdragon, petunia, etc., which are grown indoors under lights, must be hardened off before being exposed to full sun in the garden. Begin this toughening process by putting the plants outdoors for a few hours daily, gradually increasing the length of exposure. A cold frame is a big help at this time.

The experimental setup I put in my cellar consisted of two 40-watt tubes and a timer which turns the lights on about 8 A.M. and shuts them off at 9 P.M. In this particular case appearance did not matter. All that was required was that it should be functional. An old kitchen table was set up on a low platform made with boards and sawhorses. Two pieces of lumber 2 by 4 by 48 inches were laid on the table to support four aluminum trays, each 17½ by 11½ by 1 inches, in which the pots, etc., were placed. A roll of aluminum foil provided an effective reflector. This was supported on a piece of lumber 48 by 1 by 1 inches, which was held 8 inches from the

Fluorescent lighting in author's cellar.

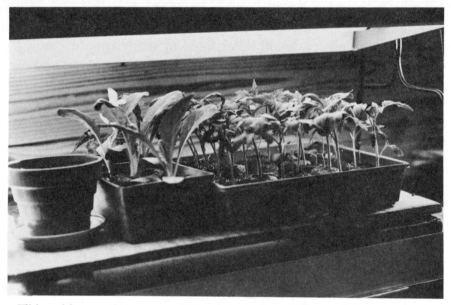

This table-top fluorescent light unit is especially useful for starting vegetable and flower garden seeds indoors in spring.

wall by two brackets. Aluminum foil 17 inches wide was cut into three lengths extending from below the trays over the bar and then over the fixture itself, thus enclosing the whole setup except for the ends, which were left open. When it was necessary to care for the plants, the foil was just thrown up over the top, and then, when the job was finished, the flaps were brought down again to the original position.

If you want something a little more fancy than the preceding, here is an arrangement from the U. S. Department of Agriculture. It is equipped with a heating unit to provide bottom heat. It is desirable, although not absolutely necessary, to maintain a temperature of the rooting medium 5° to 10° higher than that of the air. The front of the case is hinged so that it can be swung back. Camellia cuttings are in the center flat.

Another advantage of the reflector was that it helped to maintain the temperature three or four degrees higher than outside. Also the humidity was increased from 10 to 20 per cent because the evaporation of moisture was confined to some extent. This was set up in our cellar, where the temperature in winter hovered around 60° F. If yours happens to run cooler than this during the winter months, it would be desirable to install, in addition to the flourescent lights, one or two incandescent lamps, which probably would be enough to raise the temperature 10° or 15° F. Or a thermostatically controlled soil-heating cable can be installed in a tray before the growing medium is added.

This really worked out very successfully, especially with leaf cuttings of African-violet (*Saintpaulia*), which reached flowering size weeks ahead of those which were kept in my study, where the lighting was natural. There was one plant, a poinsettia, which definitely resented being in this enclosure. It was a healthy plant when it was brought in. Its flower buds were already visible, but after about three weeks it showed signs of deterioration and by Christmas was a wreck.

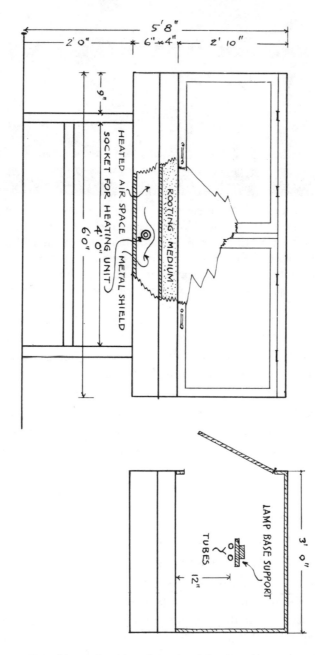

Working plan showing the false bottom to
accommodate the heating unit.

APPENDIX

PROPAGATION NOTEBOOK

KIND	ROOTED	REMARKS
		(Note: s. stands for sand; p.m. and s. stands for peat moss and sand)

Inserted May 17

Neillia sinensis	June 10	Remarkably quick rooting.

Inserted June 8

Corylopsis pauciflora	Sept. 1	No difference caused by position of cuts.
Corylopsis spicata	Sept. 1	Better in p.m. and s. with cut made between nodes.
Cytisus x *praecox*	July 21	Current-season shoots only rooted. No difference, s. and p.m. and s.
Daphne genkwa	Aug. 10	Best in sand. Root cuttings preferred.
Fothergilla major	July 22	Better in p.m. and s. with cut made between nodes.
Syringa vulgaris	Aug. 3	In s. only. No roots in p.m. and s.
Parrotia persica	Aug. 17	No difference in s. and p.m. and s. No difference in position of cut.
Polygala x *dalmaisiana*	July 27	Best in p.m. and s.
Stachyurus chinensis	Sept. 11 (30%)	Better in s.
Poncirus trifoliatus	Aug. 3	Better in p.m. and s.

Inserted June 15

Rhododendron (*Azalea*) *species*	Aug. 3	Better roots in p.m. and s.
R. 'Daviesi'	Aug. 17 (100%)	P.m. and s. best.
R. obtusum 'Hinodegiri'	Aug. 17 (100%)	In s. and p.m. and s., but better roots in latter.
Berberis verruculosa	July 24 (1 of 6)	In s. All dead in p.m. and s.
Jasminum nudiflorum	July 21	Best in p.m. and s.
Lagerstroemia (Crape-myrtle)	July 21 (100%)	Better roots in p.m. and s.
Osmanthus aquifolius	Aug. 3 (100%)	Best in s. No difference in position of cut.
Pieris japonica	Aug. 10	Heel cuttings; p.m. and s. best.

KIND	ROOTED	REMARKS
Salix babylonica annularis	Aug. 10 (75%)	Too soft.
Tamarix odessana	————	⎫ Did not root. Too early.
Tamarix pentandra	————	⎭ (See Inserted July 7)
Taxus brevifolia	Sept. 26 (100%)	Rooted more quickly in s. Best results in p.m. and s.
Viburnum carlesii	Aug. 3	P.m. and s. best. Those in s. alive but unrooted Sept. 12.

Inserted June 19

Buxus sempervirens	Sept. 26 (100%)	Both old and young wood. Best in s.
Cotoneaster horizontalis	Aug. 3 (100%)	Best in p.m. and s.
Thuja occidentalis	Sept. 26 (50%)	Best in p.m. and s.
Thuja orientalis	Sept. 26 (100%)	Best in p.m. and s.

Inserted June 28

Viburnum plicatum tomentosum	Aug. 10	Did better than June 15 insertions. 100% rooted in s. Heel cuttings; or cut through "rings" on stem better than below nodes.

Inserted July 7

Corylopsis willmottiae	Sept. 11 (50%)	Best in p.m. and s.
Magnolia stellata	Sept. 26 (20%)	P.m. and s. and cut between nodes.
Tamarix odessana	July 27	⎫ P.m. and s. Earlier insertions did
Tamarix parviflora	July 27	⎭ not root.
Viburnum utile	Aug. 17 (100%)	In s.

Inserted July 17

Daphne cneorum	Sept. 1 (100%)	In s.

Inserted July 25

Rhododendron obtusum 'Hinodegiri'	Sept. 6 (100%)	Two-year wood rooted as easily as that of current season.
Berberis julianae	Sept. 5 (100%)	No difference s. and p.m. and s. Cut between nodes better.
Berberis verruculosa	Sept. 5 (100%)	P.m. and s. best—slower in s. Cut between nodes best.
Mahoberberis x *neubertii*	Sept. 5	Same result as June 15.
Mahonia aquifolium	Sept. 5 (100%)	Cut between nodes. P.m. and s. gave rooting twenty days earlier.
Viburnum tomentosum	Sept. 1 (100%)	P.m. and s. Better than earlier insertions.

KIND	ROOTED	REMARKS

Inserted July 28

Chamaecyparis pisifera 'Plumosa'	Sept. 26 (50%)	Roots in p.m. and s. two weeks earlier than in s.

Inserted August 4

Rosa chinensis minima	Sept. 1 (100%)	No difference in s. and p.m. and s. or position of cuts.
Rosa chinensis viridiflora	Sept. 6 (100%)	P.m. and s. best.

Inserted August 10

Berberis verruculosa	Oct. 3	100% rooted in p.m. and s. Those in s. alive but not rooted.
Hydrangea (Pee Gee)	Aug. 31	Well rooted in p.m. and s. Those in s. barely started.
Pieris japonica	Sept. 26	100% p.m. and s.

A study of the preceding table will show that there are queer variations in the behavior of cuttings (notice especially *Berberis verruculosa*), caused by time of insertion, position of cuts, and kind of rooting medium, all of which tend to make the whole process of absorbing interest.

UNDERSTOCKS AND SCIONS

It is necessary for success in grafting that the understock and the scion be compatible, which means that they must be related botanically. Following is a list of plants and the understocks on which they may be grafted. The understocks usually are raised from seeds, occasionally from cuttings and layers.

Apple (dwarf)	'Malling 9'
Camellia	C. japonica
Catalpa	C. speciosa
Chaenomeles (Flowering Quince)	C. japonica
Chaenomeles (Quince)	Common Quince
Citrus (Orange, Grapefruit, Tangerine)	Sour Orange and Pomelo
Cornus (Flowering Dogwood)	C. florida
Malus (Apple)	Common Varieties French Crab Apple
Malus (Flowering Crab Apple)	Common Apple
Pear (dwarf)	Quince
Picea (Spruce)	Norway Spruce

Pinus (Pine)	
2-needle	*P. sylvestris*
5-needle	*P. strobus*
Prunus (Apricot)	Common Apricot
Prunus (Cherry-sour)	Mahaleb
Prunus (Cherry-sweet)	Black Mazzard
Prunus (Peach)	Wild from Kentucky and Tennessee
Prunus (Plum)	Myrobalan Plum
Pyracantha (Firethorn)	*Crataegus*
Pyrus (Pear)	Chinese or Japanese sand pear
Rhododendron	*R. ponticum*
Rose	*Rosa multiflora* and others
Syringa—Lilac	Privet, Lilac
Taxus—Yew	*T. cuspidata*

PLANTS WHICH ROOTED BY AIR-LAYERING[1]

	NUMBER TRIED	NUMBER ROOTED		NUMBER TRIED	NUMBER ROOTED
*Abeliophyllum			*Acer saccharum*		
distichum	1	1	'Temple's Upright'	8	1
Abeliophyllum			*Aesculus* x *carnea*	4	2
distichum	4	2	*Aesculus*		
Acer barbinerve	6	4	*hippocastanum*	6	2
Acer capillipes	3	2	*Aesculus*		
Acer circinatum	5	3	*hippocastanum*		
Acer cissifolium	1	1	'Umbraculifera'	4	1
Acer durettii	3	1	*Albizia julibrissin		
Acer ginnala	2	1	rosea	10	5
*Acer griseum	3	1	*Betula aurata*	3	1
Acer griseum	3	1	*Betula fontinalis*	4	2
Acer grosseri hessii	4	3	*Carya tomentosa*	6	1
Acer palmatum			*Castanea mollissima*	10	1
dissectum	4	1	*Catalpa bignonioides*	4	4
Acer pensylvanicum	6	6	*Catalpa bungei*	4	1
*Acer platanoides	6	4	*Catalpa speciosa*	4	2
Acer platanoides	4	1	*Cercis chinensis*	4	2
Acer platanoides			*Cladrastis platycarpa*	3	2
'Globosum'	4	3	*Clethra barbinervis*	3	2
Acer platanoides			*Cornus alba		
'Nanum'	2	2	'Sibirica'	3	3
Acer platanoides			*Cornus florida*		
var.	4	4	'Rubra'	9	8

* Layers made in 1950, all others in 1951.

[1] Reprinted from THE ARNOLD ARBORETUM GARDEN BOOK, with permission of Arnold Arboretum, Boston, Mass.

	NUMBER TRIED	NUMBER ROOTED
Corylopsis glabrescens	4	3
Corylopsis spicata	4	3
Corylus chinensis	4	1
*Cotinus coggygria purpureus	4	4
Cotinus coggygria purpureus	11	3
Cotoneaster foveolata	4	4
Cotoneaster horizontalis	4	4
Crataegus monogyna 'Stricta'	6	1
Crataegus pinnatifida major	6	1
Cytisus x praecox	4	1
Cytisus supinus	4	4
Davidia involucrata vilmorinii	15	5
Diospyros lotus	4	2
Enkianthus campanulatus	5	3
*Forsythia 'Arnold Dwarf'	3	3
Franklinia alatamaha	4	4
Ginkgo biloba	1	1
Halesia carolina	6	6
Halesia monticola rosea	4	4
Hedera helix baltica	4	4
*Hibiscus syriacus	1	1
Hippophae rhamnoides	6	3
Ilex crenata 'Convexa'	4	4
Ilex glabra	6	6
Ilex montana macropoda	1	1
Ilex verticillata	3	1
Indigofera amblyantha	4	3
Koelreuteria paniculata	8	8
Laburnum anagyroides	4	3
Laburnum x watereri	8	4
*Ligustrum ibota 'Aureum'	3	3
*Ligustrum ovalifolium	4	4

	NUMBER TRIED	NUMBER ROOTED
Ligustrum vulgare 'Buxifolium'	4	4
Lonicera maackii	4	2
Maackia amurensis	4	4
Magnolia denudata	10	1
Magnolia x soulangiana 'Alexandrina'	4	1
Malus atrosanguinea	5	4
Malus 'Dorothea'	5	5
Malus floribunda	6	1
Malus halliana spontanea	5	1
Malus 'McIntosh'	3	2
Malus x micromalus	7	2
Malus prunifolia rinkii	6	3
Malus purpurea	4	3
Malus sargentii 'Rosea'	6	3
Malus spectabilis	6	5
Malus sublobata	6	6
Malus sylvestris astracanica	4	1
Malus 'Wabiskaw'	4	2
Morus alba 'Pendula'	4	2
Orixa japonica	4	4
Osmaronia cerasiformis	4	2
*Populus alba nivea	2	2
Prunus juddii	1	1
Prunus maackii	2	2
Prunus serrulata 'Amanogawa'	4	3
Prunus serrulata 'Gyoiko'	2	1
Prunus serrulata 'Kwanzan'	6	2
Prunus yedoensis 'Taizanfukun'	4	4
Ptelea trifoliata aurea	6	2
*Rhododendron 'Dr. Charles Baumann'	7	5
*Rhododendron 'Josephine Klinger'	9	8
Salix caprea	7	7
Styrax japonica	7	2
Symplocos paniculata	6	3

	NUMBER TRIED	NUMBER ROOTED		NUMBER TRIED	NUMBER ROOTED
Syringa amurensis japonica	6	1	*Viburnum opulus	2	2
Syringa x *prestoniae* 'Lucetta'	4	3	*Viburnum rhytidophyllum*	4	1
Syringa x *prestoniae* 'Paulina'	5	3	*Viburnum rufidulum*	6	1
Syringa villosa	5	3	*Viburnum sargentii	6	5
Syringa vulgaris vars.	35	9	*Viburnum sargentii*	3	3
Tamarix pentandra	2	2	*Viburnum sargentii flavum*	4	4
Taxus cuspidata 'Nana'	3	1	*Viburnum setigerum aurantiacum*	8	3
Tilia cordata	5	3	*Viburnum sieboldii*	4	1
Tilia platyphyllos 'Fastigiata'	8	2	*Wisteria floribunda* 'Violacea-plena'	4	2
Tsuga canadensis	3	3	*Wisteria floribunda* 'Longissima alba'	4	2
Ulmus carpinifolia 'Koopmannii'	6	1	*Wisteria floribunda* 'Naga Noda'	4	4
Ulmus carpinifolia 'Sarniensis'	4	1	*Wisteria formosa*	4	2
Ulmus glabra	4	1	*Wisteria macrostachya*	4	4
Vaccinium corymbosum 'Jersey'	3	2	*Wisteria sinensis*	15	1
Viburnum carlesii	4	1	*Wisteria sinensis* hybrid	2	2
Viburnum dilatatum	4	1	*Wisteria venusta*	3	1
Viburnum juddii	4	1	*Zelkova serrata*	6	6
			Zelkova sinica	6	3

PLANTS WHICH FAILED TO ROOT BY AIR-LAYERING

	NUMBER OF LAYERS TRIED		NUMBER OF LAYERS TRIED
Abies homolepis	6	*Carpinus cordata*	4
Acer campestre	6	*Carpinus orientalis*	4
Acer palmatum 'Atropurpureum'	1	*Carya glabra*	6
Acer palmatum 'Lutescens'	2	*Carya* x *laneyi*	2
Acer platanoides	6	*Carya ovata*	4
Acer shirasawanum	3	*Carya schneckii*	4
Acer tataricum	6	*Castanea dentata*	4
Aesculus x *carnea* 'Plantierensis'	5	*Catalpa fargesii*	6
Aesculus discolor mollis	4	*Cercis canadensis* 'Alba'	8
Aesculus glabra leucodermis	4	*Chaenomeles sinensis*	4
Albizia julibrissin rosea	16	*Chionanthus retusa*	4
Amelanchier canadensis	2	*Chionanthus virginicus*	2
x *Amelasorbus jackii*	4	*Cladrastis lutea*	4
Betula jacquemontii	3	*Cornus florida*	4

	NUMBER OF LAYERS TRIED		NUMBER OF LAYERS TRIED
Cornus mas	8	*Malus ioensis* 'Plena'	1
Cornus officinalis	4	*Malus x robusta*	4
Corylus avellana 'Contorta'	3	*Malus sargentii*	6
Corylus avellana 'Fusco-rubra'	4	*Parrotia persica*	4
Corylus chinensis	5	*Phellodendron amurense*	9
Cotinus obovatus	5	*Phellodendron chinense*	4
Crataegus arnoldiana	4	*Phellodendron sachalinense*	6
Crataegus coccinioides	6	*Photinia villosa*	4
Crataegus x lavallei	4	*Pinus bungeana*	5
Crataegus monogyna 'Inermis'	2	*Prunus maritima* 'Eastham'	4
Crataegus monogyna 'Versicolor'	3	*Prunus maritima* 'Raribank'	4
Crataegus nitida	5	*Prunus serrulata*	5
Crataegus pruinosa	6	*Pseudolarix amabilis*	4
Crataegus punctata	6	*Quercus bebbiana*	4
Crataegus succulenta	3	*Quercus bicolor*	6
Diospyros virginiana	4	*Quercus dentata*	4
Eucommia ulmoides	4	*Quercus falcata*	3
Evodia daniellii	4	*Quercus marilandica*	5
Fagus grandifolia	2	*Quercus mongolica*	5
Fagus sylvatica 'Atropunicea'	3	*Quercus robur*	6
Fagus sylvatica 'Pendula'	4	*Quercus robur* 'Argenteo	
Fothergilla monticola	2	Marginata'	4
Fraxinus chinensis	6	*Quercus runcinata*	6
Fraxinus pennsylvanica	4	*Quercus stellata*	5
Gleditsia triacanthos	6	*Quercus variabilis*	4
Gleditsia triacanthos 'Inermis'	4	*Rhododendron* 'Mrs. C. S.	
Hamamelis mollis	8	Sargent'	4
*Hamamelis mollis	3	*Rhododendron* 'Purpureum	
Juglans cinerea	6	Grandiflorum'	2
Juglans nigra	9	*Sophora japonica*	4
*Juglans nigra 'Laciniata'	4	*Sorbus alnifolia*	4
Juglans nigra 'Laciniata'	5	*Sorbus aria*	4
Kalmia latifolia	5	*Sorbus aucuparia*	8
Kalmia latifolia polypetala	7	*Sorbus x latifolia*	6
Kalopanax pictus	4	*Sorbus thuringiaca*	4
Lindera benzoin	3	*Syringa pubescens*	4
Magnolia fraseri	6	*Taxus cuspidata expansa*	4
Magnolia x loebneri	6	*Thuja standishii*	5
*Magnolia stellata rosea	4	*Tilia americana* 'Fastigiata'	4
Magnolia stellata rosea	6	*Tilia platyphyllos*	6
Magnolia virginiana	4	*Tsuga caroliniana*	4
Malus 'Arrow'	3	*Ulmus carpinifolia* 'Dampieri'	4
Malus baccata	5	*Ulmus carpinifolia*	
Malus 'Bob White'	4	'Umbraculifera'	6
Malus brevipes	1	*Ulmus carpinifolia* 'Wredei'	4
Malus coronaria 'Charlottae'	3	*Ulmus plottii*	4
Malus florentina	5	*Vaccinium corymbosum*	
Malus glabrata	4	'Harding'	2
Malus hupehensis	5	*Wisteria* 'Jako'	2

INDEX

ACKNOWLEDGEMENTS

The majority of photographs were taken by Montague Free, *The Home Garden*, Malcolm R. Kinney, The Munson Studio, and Steenson & Baker except for the following (page positions are indicated by u for upper and l for lower):

American Orchid Society, Botanical Museum of Harvard University, Cambridge, Massachusetts 02138: 59–63.
George Bishop: 20.
Brooklyn Botanic Garden: 14, 15, 23u, 83, 114l, 119, 120, 122, 176l, 221.
Dennis Brown: 32u, 194, 196–201.
Marjorie J. Deitz: 2, 121, 141, 190l, 234u, 236l.
Gottscho-Schleisner: 92, 93, 123u, 236u.
Rod McLellan Company: 33.
Margaret Perry: 183–85.
Maud H. Purdy: 125 (drawings).
Seabrook Farms: 186, 187.
Henry T. Skinner: 127.
Richard A. Smith: 70, 71.
George Taloumis: 84, 85l.
Walter E. Thwing: 172–76.
United States Department of Agriculture: 26, 32l, 35l, 159l, 160u, 161l, 229, 230, 231, 237.